Human Activities and the Tropical Rainforest

The GeoJournal Library

Volume 44

The titles published in this series are listed at the end of this volume.

Human Activities and the Tropical Rainforest

Past, Present and Possible Future

edited by

BERNARD K. MALONEY

The Palaeoecology Centre,
The Queen's University,
Belfast, Northern Ireland

KLUWER ACADEMIC PUBLISHERS
DORDRECHT / BOSTON / LONDON

A C.I.P. Catalogue record for this book is available from the Library of Congress.

ISBN 0-7923-4858-3

Published by Kluwer Academic Publishers,
P.O. Box 17, 3300 AA Dordrecht, The Netherlands.

Sold and distributed in the U.S.A. and Canada
by Kluwer Academic Publishers,
101 Philip Drive, Norwell, MA 02061, U.S.A.

In all other countries, sold and distributed
by Kluwer Academic Publishers Group,
P.O. Box 322, 3300 AH Dordrecht, The Netherlands.

Printed on acid-free paper

Printed in the Netherlands

TABLE OF CONTENTS

Dr. Kenneth McGuffie,
Centre for Climatic Impacts,
McQuarie University, NSW 2109,
Australia.

Professor Ann Henderson-Sellers,
Royal Melbourne Institute of
Technology, Melbourne, Victoria
3001, Australia.

Dr.Huqiang Zhang,
Bureau of Meteorology Research
Centre, P.O. Box 1289K, Melbourne,
VIC 3001, Australia.

Dr. Bernard K. Maloney,
Palaeoecology Centre, The Queen's
University, Belfast.

PREFACE

This book arises out of a one day conference organised by the Development Studies Committee at The Queen's University, Belfast, in February 1993 jointly with the Geographical Association (Belfast Branch). The editor was a committee member of both organisations, and the Chairperson of the Development Studies Conference Organising Committee. Over 300 people attended the conference sessions and Professor Sir Ghillean Prance delivered the keynote lecture to a large audience.

So, the conference was a success, and discussion among the speakers, and with Dr. J. McAdam, the Chairperson of the Development Studies Committee, led to agreement that we should (Maloney should!) approach other possible contributors with a view to publishing a multi-disciplinary, but thematic, account of human activities the past, present and future development of the tropical rainforest. This might seem to be an easy task, but, because many academics are so very overworked, it took over a year to find suitable people who were prepared to contribute to the book, then a publisher had to be sought. Following refereeing of the proposal, Kluwer Academic kindly agreed to publish this book as one of their Geo-Journal series, and the editor wishes to thank them for doing so, and, in particular, Ms Petra van Steenbergen for her enthusiastic co-operation through the time period this volume has taken to produce.

Special care was taken to make sure that there was cover of the physical background of the rainforest: its geomorphology and soils and how these impact upon natural and anthropogenic change, the long term and more recent history of human activity in the forest, and an assessment of reafforestation, both socio-economically, and in terms of reduced biodiversity. The topics interrelate the past and present evolution of the rainforest, but the past and present influence the future, and the possible future of the rainforest is considered in more detail using a modelling approach. Some of these themes have, inevitably, been covered, or partly covered, elsewhere before while others have not. What has been assembled is a collection of essays which is unique in its range, and the authors feel that a broad brush approach, with some more specific examples, is necessary to put the situation in perspective. Some people might argue that a book with a title such as the one given to this portends a 'doomsday' book. The editor would disagree: it is simply a statement of what has occurred in the distant past, and what was very important in terms of largely non-destructive use of forested areas, of the technological developments that took place, which inevitably brought about forest clearance during prehistory, what happened in colonial times, the perspective of some remaining tribal societies, where we are at at present, and what the possible future is for the remaining rainforest, all set against the backdrop of the fragility of tropical physical environments.

There is no intention to preach to countries that still have rainforests about how they should conserve them. This is **not** a book about rainforest politics. Indeed it would be arrogant of us in the so-called First World to write telling the people of the so-called Third World what they should do with the forest, especially as eco-tourism from the rich countries in the future may, if not carefully controlled, pose as much a problem to the maintenance of those areas of forest which have been conserved as loggers and miners have to those which have not been preserved. We do not expect so-called Third World countries to tell us how we should conserve our environment and are in no moral position to tell them what they should do, especially as we have made such a bad job of managing the environment in so many areas. We would be better advised to say 'look what a mess we have made, if we can help you to avoid similar disasters, and you want that help, ask us and we will advise as best we can, but we have no right to tell you what to do.' This book is concerned with what was, is, and what could be. It does not aim to present a First World opinion of what ought to be. It may be claimed, 'but the rainforest belongs to the world'. That might be so, but the land which is occupies and the minerals and ores beneath that land does not, therefore a more pragmatic approach to conserving what remains of the rainforest is advisable.

Because of the wide-ranging nature of the contributions, it is hoped that this book will appeal across the range of inter-disciplinary divides: to physical geographers, archaeologists, foresters, and agronomists to name but a few. Indeed it is also hoped that its content may appeal to the more general reader, if only to dip into.

Bernard Maloney   Belfast, 1997

ACKNOWLEDGEMENTS

Bernard Maloney would like to thank Dr. J. Mallory, Department of Archaeology, Queen's University, Belfast, for critical reading of Chapter 4, Ms. Maura Pringle of the School of Geosciences for preparing many of the text figures and Dr. F.G. McCormac for many discussions and help with computing problems. Most all, however, he would like to thank Dr. Lisa Kealhofer, Department of Anthropology, College of William & Mary, Williamsburg, Virginia, whose e-mail messages kept him going when the will to finish the task of writing and editing was wilting.

An earlier version of Roy Ellen's paper was presented at the 1993 British Association (Section E) meeting held at Keele University. He thanks Professor Ian Douglas, Department of Geography, University of Manchester, for the invitation to speak at that meeting. Writing and subsequent revison of the paper was supported by ESRC grants R000 23 3088 (The ecology and ethnobiology of human-rainforest interaction in Brunei: a Dusun case study) and R000 236082 (Deforestation and forest knowledge in south central Seram, eastern Indonesia), in association with the EC funded programme, *Avenir des Peuples des Forets Tropicales* (APFT).

The work of McGuffie *et al.* was partly funded by grants from the Model Evaluation Consortium for Climatic Assessment, the National Greenhouse Advisory Committee of the Department of the Environment, Sport and Territories, and by the Australian Research Council. Their chapter is contribution number 95/45 of the Climatic Impacts Centre, McQuarie University.

Permission has been gratefully received from the following to reproduce copyright material: Geological Society, London (Thomas, Fig. 1), Crown Agents and A.T. Grove (Thomas, Fig. 5), Prentice-Hall Inc. (Furley, Fig. 1b) and the University of Oklahoma Press (Furley, Fig. 3). Efforts to trace the copyright ownership of some material have proved impossible and this opportunity is taken to offer apologies to any copyright holders whose rights may have been unwittingly infringed.

CONTRIBUTORS

Christopher J. Barrow Centre for Development Studies, University College of Swansea, Swansea SA2 8PP, Wales.

Roy F. Ellen Department of Sociology and Social Anthropology, Eliot College, University of Kent at Canterbury, Canterbury, Kent CT2 7NS, England.

Alastair Fraser UK-Indonesia Tropical Forest Management Project, Manggala Wanabakti Building, Jalan Gatot Subroto, Jakarta 10270, Indonesia.

Peter A. Furley Department of Geography, The University of Edinburgh, Edinburgh EH8 9XP, Scotland.

Ann Henderson-Sellers Royal Melbourne Institute of Technology, Melbourne, Victoria 3001, Australia.

Bernard K. Maloney Palaeoecology Centre, The Queen's University, Belfast BT9 6AX, Northern Ireland.

Kenneth McGuffie Department of Applied Physics, University of Technology, Sydney, New South Wales 2007, Australia.

Stephen Nortcliff Department of Soil Science, The University of Reading, Reading RG6 6DW, England.

Sir Ghillean T. Prance Royal Botanic Gardens, Kew, Richmond, Surrey TW9 3AB, England.

Michael F. Thomas Department of Environmental Science, University of Stirling, Stirling FK9 4LA, Scotland.

Huqiang Zhang Bureau of Meteorology Research Centre, P.O. Box 1289K, Melbourne, VIC 3001, Australia.

FOREWORD

I am delighted that the exciting one day rainforest conference, with relatively few speakers, at which I delivered the keynote address, has been expanded to produce this volume on the much neglected topic of human activities in the tropical rainforest. The coverage of this book is both broad geographically and in the range of disciplines involved. It also covers a wide time span from the geological formation of the rainforest landscape over the ages to the future effects of climatic change and the impact of deforestation upon climatic change.

This is not the usual symposium volume with a number of disconnected chapters. It has been carefully put together to develop the topic of human interaction with the rainforest and to provide the basic information about rainforest, in order to understand the implications by our species. By introducing and concluding the volume editor Bernard Maloney brings together well the various topics addressed here.

Tropical rainforest covers only about 7 per cent of the land surface of the world, yet it harbours at least fifty per cent of the biological species. Human activity within rainforest areas will inevitably cause the loss of many of these species. Some of these species are well known and have already been of considerable benefit to human beings, for example rubber, cacau, bananas, papayas, oil palms and vanilla to name but a few important crops which have originated in rainforest areas. A large number of medicines which we use come from tropical plants and even more plants are used as medicines by local peoples throughout the world. The concern about the loss of tropical forests should not only be for those plants and other orgamisms which we already use, but for the numerous undiscovered medicines, agrochemicals, oils, fruits and fibres. Of particular concern is the preservation of the wild relatives of known crops, because survival of a crop plant often depends on genetic material from wild species. Disease resistance, crop improvement or adapting a crop for different climates usually relies upon genetic material in their wild relatives. McGuffie *et al.* demonstrate that removal of tropical rainforest causes change in local climate. We also need the biodiversity to adapt to broader worldwide climate change. *Eucalyptus* has a role in sustainable land use, particularly in areas which have already been degraded, but it would be a tragedy if species diverse forest is all replaced by a monoculture of a single species or genus.

Indigenous people have lived in the rainforests of the world for many generations and several chapters in this book show that they have not necessarily destroyed all the species. Even in the Mayan areas of MesoAmerica where there has been intensive land use over a long period there is a rich biota and Furley shows that the Selva Maya remains one of the most important and biologically rich places in the region. The challenge before us today is to halt the wanton destruction and use the land in more sustainable ways that does not eliminate the species. This volume gives many hints towards sustainability. It even has a chapter on the soil which is indeed a much neglected component of these ecosystems, yet it is often the understanding and management of soil that is the key to successful land use.

The papers in this book show that there is a great deal of research still to be done to achieve sustainability, whether it be the basic taxonomic classification of organisms or applied work on sustainable forestry and agriculture.   The gaps in knowledge shown by the authors in this volume are probably the most important aspect because they should stimulate further research to help us to respond to the needs and to seek solutions that balance wise land use with conservation of species, and that helps us to understand the effects of long term change.

Ghillean Prance, FRS
Director,
Royal Botanic Gardens,
Kew.

1. INTRODUCTION

Bernard K. Maloney

1. Definition of the tropical rainforest

The title of this book is 'Human activities and the tropical rainforest: past, present and possible future'. Rainforests of varying types are found in the humid and seasonally dry areas of the tropics and the mountains which they encompass but not all of these regions are naturally forested. There are, for instance, natural grasslands along part of the Orinoco River in Venezuela.

Strictly defined the term 'tropics' refers to the lands and islands between the Tropic of Cancer and Capricorn (23.5° N and 23.5° S of the equator), areas where the sun can lie at zenith (Reading *et al.* 1995). These are areas which receive large amounts of solar radiation during the year and therefore have minimal seasonal temperature variations although the mountains (Troll 1959) may have a 'thermal-diurnal' climate, in effect summer every day and winter every night. It has often been said therefore that the most important climatic parameter in the tropics is the fluctuation in precipitation throughout the year. Some areas of East Africa have two wet seasons and two dry seasons, while most of the drier areas of Southeast Asia have one dry season differing in length northwards on the mainland and eastwards in the Indonesian islands. However, the tropics between five degrees north and south of the equator are everwet and it is there, in the lowlands, that the rainforest proper is to be found.

Rainforests of different types are present naturally where mean annual rainfall is above about 2000 mm. Where the precipitation is lower than this various forms of dry forest, e.g. the dry evergreen forests of Myanmar and Thailand occur. Writers generally label these too (Whitmore 1975, 1990; Jacobs 1987) as rainforest. It is also usual to include the seralmangrove and peat swamp formations of the lowland coastal areas and sub-montane, lower montane and upper montane forests of the highlands. Indeed it is from areas where these occur, or formerly occurred, that most of the information on the history of the rainforest from the Tertiary through to the present and information on the past usage of the forest by peoples in Africa, Central and South America and Southeast Asia has been derived, primarily through the analysis of pollen deposited and preserved in lakes, peat bogs, estuarine and marine sediments (cf. Maloney, this volume) but increasingly (cf. Thompson 1994) from macrofossils recovered by careful sieving of materials from archaeological excavations.

Reading *et al.* (1995) discussed the problem of definition of the humid tropics in detail, and it is not intended to repeat the full account of what they wrote, but since the rainforests with which we are concerned are found in the humid tropics their conclusions merit a summary here.

They stated that the astronomical definition was too rigid for most geographical and ecological purposes and that botanists required a definition based on vegetation assemblages, climatologists on the basis of the prevalence of specific atmospheric conditions and geomorphologists prefer a definition derived from the intensity or magnitude of the physical processes which operate. Unfortunately the boundaries which most please each of the disciplines do not coincide.

1

B.K. Maloney (ed.), Human Activities and the Tropical Rainforest, 1-15.
© *1998 Kluwer Academic Publishers. Printed in the Netherlands.*

As already indicated, rainforests vary in their nature with latitude and altitude. Latitude is also an important factor in determination of climatic conditions and climate. Apart from its direct or indirect influence on the sort of rainforest which occurs, or once occurred, it affects the nature of landform processes, the soils, agriculture and economic development. In short, the very things which this book aims to discuss from the point of view of human impact on the rain forest.

There have been a number of attempts to delineate the extent of the tropics using climatic parameters but climatic definitions fail because climatic is more dynamic than the other factors under consideration. Not only has climate changed considerably over the time period since theTertiary in particular but there are air flows in to and out of the tropics. Definitions based on the dynamic nature of climate are, therefore, unsatisfactory. Others are based on more constant factors, e.g. temperature, perhaps the most constant climatic factor, and rainfall, which is less constant in those areas of Africa and, particularly, Asia which are subjected to monsoonal circulations. Definition of what a monsoon is is difficult enough. Is it a wind, a precipitation regime or a circulation system? There is disagreement among climatologists. If precipitation is the most important climatic statistic in the tropics, it is this acting through the process of water availability (Savage *et al.* 1982) which should be the lynch pin of classification but, as Reading *et al.* (1995) point out, the units used and the limits imposed vary massively from scheme to scheme. There are meteorological classifications and hydro-meteorological classifications but none of them are entirely satisfactory, partly because of the range of altitude present in tropical areas which, in Southeast Asia alone includes land at sea level to that above 4000 m (Mount Kinabalu in Sabah, which was glaciated during the Late Quaternary).

Koppen (1936) distinguished a tropical rainforest climate (Af), a tropical monsoon forest climate (Aw) and a tropical savanna climate. Rainforest as commonly defined only occurs in the first two of these areas but the rainforest and monsoon forest boundaries have oscillated over time with changes in climate and, more recently, possibly, changes of climate combined with the activities of people. Tricart (1972) was concerned with areas which showed humid tropical weathering patterns and profiles, and the division which he employed was one of constantly humid (the region of our rain forest *sensu stricto*), constantly humid with a short dry season (where monsoon forest would occur naturally?) and seasonally humid with a long dry season. The real problem, and one which has not been satisfactorily resolved, is definitions of the climatic limits to the range of the various rainforest types. This requires a definition based on the seasonality of rainfall.

Reading *et al.* (1995) acknowledge that the number, variation and degree of sophistication of the classification systems which they discuss can appear bewildering and conclude that a single scheme which would fit, or could be adapted to fit, the many aspects of the humid tropical environment which they discuss could not be found. As the nature of the tropical rainforest is so diverse from place to place, the same conclusion has to be drawn here. This is not an elegant approach, but it is a pragmatic one. They accept that areas could be regarded as tropical due to their consistently high receipt of solar radiation, heat and moisture (the last two reduce to sea level equivalents) and that these climatic factors give rise to distinct patterns of vegetation, animal life, soil, landform development, agriculture and economic development. This definition is vague, but it was the working definition on which their book was based and, at risk of being accused of subscribing to the reinforcement syndrome, it will be adopted here for the lack of anything better. Under this schema, as they indicate, only 4% of the large area between the tropics of Cancer and Capricorn is wet enough to support tropical rainforest, the region in Koppen's Af and Am categories.

In terms of landforms it includes the ancient shield areas of South America and Africa, the much younger regions of Southeast Asia where the continental plates are believed to have first collided in the Miocene and the volcanic islands of the Pacific which are also geologically young but which support, or supported before the Polynesian migrations, a rainforest which is depauperate compared to that of Southeast Asia and Central and South America. Indeed what may be called rainforest proper extends outside the limit of the Tropic of Cancer and is largely virgin because of the rugged topography and lack of access roads on some of the Hawaiian islands.

2. Classification of tropical rainforest vegetation

There are probably as many classifications of tropical rainforest vegetation as there are definitions of the humid tropics and as with tropical soil classifications, these are often difficult to correlate. There are floristic classifications and ecological classifications. Obviously the former are more difficult to correlate from continent to continent and region to region in a single continent than the latter and since the chapters in this book examine widely varying parts of the tropics, it is more appropriate to consider the ecological classifications which are largely based on physiognomy and life-form types, essentially a phytosociological approach, than the floristic ones. Reading *et al.* (1995) have outlined some of the physiognomic classifications relating vegetation type to environmental factors and state that the most quantitative classification is that of Holdridge (1967) which relates life-form zones to climatic factors. In his classification each bioclimatic zone has a characteristic climax vegetation with a distinct ecology. However, the climatic climax hypothesis is increasingly under question resulting from the findings of palynologists working in Papua-New Guinea that some taxa in montane rainforest seem to behave individualistically (Walker and Pittelkow 1981). Similar findings have not yet been reported from the Central and South American and African tropics though. A variant of this approach has been used throughout Asia by the French School at Pondicherry, India.

Holdridge (1967) defined 125 bioclimates and 38 of these are tropical with 17 of them humid tropical. The lowland tropics has a restricted range of these but the entire range of humid tropical life zones is found in the northern Andes, East Africa and some Southeast Asian and Pacific islands. Reading *et al.* (1995) claim that climate-vegetation relationships are important because they allow latitudinal and altitudinal vegetation trends to be examined simultaneously and comparisons to be made with soil and climato-geomorphological zones. They seem to accept the view of Walter (1979) that the altitudinal changes relate to temperature but this has been challenged, quite logically, by Grubb (1971, 1974) as most tropical mountains have a daily cycle of cloud build up and dispersal which results in the presence of an upper montane zone of very wet mossy forest where the trees can be festooned with lichens and, indeed, their Figure 5.1 which is entitled 'Climate-vegetation relationships in Venezuela' derived from Walter's work has cloud forest depicted upon it. It also neglects the Massenerhebung effect, the importance of mountain mass in determining the altitudinal gradient. Vegetation zones extend down to lower altitudes on lower mountains (van Steenis 1962) as a result of this. Of course none of the classifications is absolute, but this writer prefers that of Whitmore (1990) which is shown on Table 1.

Lowland tropical rainforest covers the largest area of the various vegetation types delineated because vast areas of the humid tropics consist of low-lying Precambrian shields with many large river flood plains such as those of the Amazon and Orinoco in South America and the Congo in Africa. In contrast tropical montane and tropic-alpine vegetation is mainly found in the geologically younger regions of Central America, the Caribbean islands, the Andes and Southeast Asia. The natural distribution of forest cover in the humid tropics, excluding the Pacific islands is depicted on Figure 1.

Lowland evergreen rainforest usually comprises closed canopy forest with three or four strata and canopy trees 30-60 m tall and emergents which are even taller, gregarious dominants are uncommon. It is species rich and the trees are festooned with epiphytes, woody climbers and, often stranglers but the ground vegetation is sparse, frequently consisting of tree saplings. Forests tend to vary in species composition and be higher still along waterways. This is rainforest proper. All the otherrainforest formations have a simpler structure and fewer species.

Tropical semi-evergreen rainforest is a closed forest with the emergent trees usually scattered but sometimes 45 m in height (Whitmore 1990) present in the lower Amazon and most of the African rainforest block, including all of the Zaire basin. It has an outlier in India, was probably more extensive in Southeast Asia in the past and may be the type of rainforest which occurs in Central America. Big woody climbers are abundant and about one third of the trees are deciduous and can be gregarious. Epiphytic ferns and orchids are occasional to frequent. The number of species is high, but not as high as in evergreen rainforest, and bamboos are present.

Lower heath forest, called *kerangas* in Sarawak, is found on deep lowland podzols, peat swamp forest grows on lowland domed bogs which resemble the raised bogs of Europe and freshwater swamp forest is present in river valleys where there is daily, monthly or seasonal flooding. These are less species rich than dryland forests but contain similar plant taxa.

Heath forest is dense and often difficult to penetrate. It has a low, uniform canopy with no trace of layering and is found on very acidic, freely drained podzols developed from sandy coastal alluvium and soils derived from weathered sandstone. Epiphytes are common, but big woody climbers are rare.

Despite the attention paid to the Sarawak heath forests (cf. Brunig 1971), the most extensive heath forests are present (Whitmore 1990) in the upper reaches of the Rio Negro and Rio Orinoco in South America. Heath forest also occurs in Kalimantan and Brunei in Southeast Asia, and Gabon, Cameroun and the Ivory Coast in west Africa.

TABLE 1. Classification of tropical rainforests

Climate	Soil water	Main soil types	Approximate altitudinal ranges (m)	Forest formation
Seasonally	Large annual deficit			Monsoon forests
	Small annual deficit			Semi-evergreen forest
Everwet	Dryland	Acrisols, ferralsols	Lowlands	Evergreen rainforest
			1200-1500	Lower montane forest
			1500-3000	Upper montane forest
			3000+	Subalpine forest to tree line
		Acrisols, podzols	Mainly lowlands	Heath forest
		Ferralsols	Mainly lowlands	Forest on limestone
		Nitosols	Mainly lowlands	Forest on ultrabasics
Brackish water	Watertable high at least	Acrisols, regosols,		Beach vegetation
		Regosols, salt-water histosols		Mangrove forest
Freshwater		Histosols	Histosols	Peat swamp forest
	More or less permanently wet	Gleysols		Freshwater swamp
	Periodically wet	Fluvisols		Freshwater periodic swamp

After Whitmore (1990: 13, Table 2.1)

Peat swamp forest is defined more by the habitat in which it is found that its structure and physiognomy. The number of species present, and their height, is greatest at the edges of these domed bogs which may be up to 20 km across. Up to six types of forest occur and the outermost trees can be up to 50 m high and are regarded as a valuable source of timber. These swamps formed as sea levels rose to heights above the present datum in the middle Holocene and peat growth continues today. The peat at the centre of the dome is up to 13 m deep (Muller 1972) and very acidic as its nutrient supply is derived from rainfall.

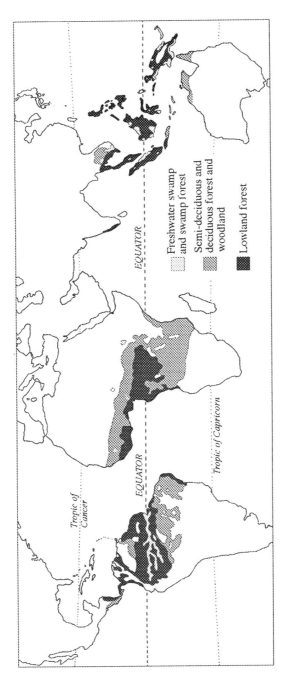

FIGURE 1. Distribution of the tropical rainforest.

Freshwater swamp forest extends up to 600 km inland along the Amazon, which floods annually. There are swamps which are permanently under freshwater, *igapo*, along silt-laden, white water rivers, and those which are flooded periodically, *varzea*, fringing black water rivers. The species present along eutrophic white water rivers differ from those found along black water rivers. Freshwater swamp forest was widespread in the alluvial plains of Asia (cf. Takaya 1987) but has mostly been cleared for wet rice cultivation and disturbed swamp forests occupy about one third of the Zaire basin (Whitmore 1990).

The other main vegetation formation present in the lowlands as a narrow, zoned, species poor forest at the coast but extending inland along tidal estuaries is mangrove swamp which Whitmore groups with beach vegetation and brackish water forest.

These plant formations give way to lower montane forest above an altitude of about 1000 m on larger mountains but upper montane forest abuts directly on lowland rain forest on smaller peaks. Lower montane forest has trees 10-30 m tall (Reading *et al.* 1995) and it is considerably less species rich than lowland rain forest but still has climbers and epiphytes, although there are more epiphytes than climbers. There are usually two strata and there are far fewer emergents. On poor soils there can be single species dominance.

Upper montane forest has a single stratum which can be of low height but may reach 20 m depending upon the altitude. It lacks emergents and has no climbers but epiphytes can be present. Again, there may be single species dominance.

Montane rainforest occurs in Cameroun and the eastern fringe of the Zaire river basin, otherwise it is rare in Africa while seasonally dry tropical montane woody vegetation is common (Whitmore 1990).

Tropic-alpine communities include heathlands, fern meadow, moss tundra and grassland, called *paramo* in the Andes (Whitmore 1990). They can include as shrubs species found as trees lower down the mountains and frequently have a high proportion of endemic species due to their isolation (Reading *et al.* 199).

3. Why is the tropical rainforest important?

The tropical rainforest is important in many different ways. It is important to the tribal peoples of the forest, as Ellen (this volume) indicates because it is their home and the source not only of their livelihood (of course it does provide them with food, materials to build shelters, boats, other artifacts) but their whole cosmology, and social and political networks also revolve around the forest. So, it is not only home, it is also a sacred place, but now it is increasingly being removed by other people who do not have their origin in the forest, who do not appreciate the way of life of the tribal peoples, and who have no strong attachment to the forest. Such people include those who have voluntarily resettled, or who have been involuntarily resettled, indigenous, and more often exogenous exploitative timber, plantation and mining companies. There has been increasing friction between so-called First and Third World countries about the rate of forest clearance, partly in relation to the controversy over global warming as the tropicalrainforest is a CO_2 sink, and because of the need to conserve what the former would regard as a world resource, but which the latter regard as their land with which they can do what they want without outside interference. In some ways the tribal peoples are 'piggy in the middle'. Whose land is it anyway? A socialist, with a small 's' view would possibly be that the land is nobodies and everybody's! It is not proposed to persue these arguments further here, more than enough has been written, and continues to be written, about them elsewhere.

However, what is sometimes neglected by the protagonists is that the remaining tropical rainforests are not just the home of the an enormous number of higher plant species and a declining number of tribal cultures but of animals and insects, of soil fauna and flora, and of fungi, and, and these perhaps even more than the higher plants, have not been adequately scientifically investigated. Some of the lower life forms as well as the flowering plants may prove to be very useful to people and the tropical rainforest houses a vast reservoir of material for future genetic engineering.

Social anthropologists are increasingly studying ethnobotany but there is scant purpose in obtaining this data about traditional, particularly medicinal uses, of forest plants if they are to be rendered extinct by widespread forest clearance. It just does not make economic sense at a national level to extend crop cultivation into lands which, at best, are marginal to agriculture, but it is frequently local and not national factors which are responsible for such expansion and these are manifold and complex in nature. It makes sense to intensify agriculture on the better land, land which has already been used for centuries, in some cases thousands of years.

Indonesia is, for instance, doing both. In the early days of transmigration from Java, Bali and Madura into the so-called Outer Islands of Sumatra, Kalimantan and Sulawesi, a process which began on a small scale in South Sumatra under Dutch colonial rule in 1905 (Sumintardja 1976), no consideration was made of the limitations imposed by the environment (Secrett 1986). Peasant farmers from Java moved to Sumatra and other so-called 'Outer Islands' and became peasant farmers there, struggling in an alien environment a long way from home (Babcock and Cummings 1985; Budiardjo 1986; Otten 1986; Anon 1989). They were no better off than they had been in Java. That has all changed in recent years (Harrison 1977; Conway and McCauley 1983). Now proper land surveys are carried out, tree crops are being planted as well as rice, and there is increased agricultural intensification in west Java. Indonesia is lucky: it has oil revenues which have, in recent years, been used to promote development. Many other developing countries with spiralling population growth are not so fortunate with spiralling population growth are not so fortunate. They do not have the resources to fund an industrial revolution to absorb the surplus rural labour supply. They are starting from a different base level to that which Britain, for instance started with, and the only lands which they can 'colonise' to increase food supplies without increasing inputs of synthetic fertilizers, which they often have to import at a cost in terms of foreign exchange, are those which they regard as under used, and those are usually the lands on which tropical rain forest is to be found.

It is easy to see development of areas under the tropical rainforest from an altruistic western concern with conservation but it is also easy to see it from the perspective of countries which have become land hungry because of rapidly growing populations. What is less easy for westerners to take is wanton destruction facilitated by the politicians of developing countries eager for an easy source of foreign exchange

which may, or may not, be used for rural development or industrialisation, but this too
is clearly not a one way process. Rural development of most tropical rainforest areas
was begun in an exploitative way by the various colonial powers, some more actively
than others, destruction of large areas of the remaining tropical rainforest is mainly a
result of neo-colonialism. The aim here is not expression of a political statement but
simply to state that it is clear that in very many instances rainforest destruction is the
result of western capitalism. That the developing countries are still being exploited and
that, perhaps, it will only be, in many instances, when, if ever, the recommendations of
the Brandt Report are implemented that the rate of felling will slow down. It is difficult
to envisage it ceasing.

There are countries in Southeast Asia, which is atypical of the whole of the
rainforested area, which are developing rapidly. Indonesia, Malaysia and Thailand are
good examples, with Vietnam, perhaps soon to follow (cf. Anh 1994; Devan 1994), and
there is evidence that people in some of these countries, perhaps a small minority, are
now becoming concerned about their environment, and environmental deterioration in
particular. There certainly is concern in Thailand, which has rapidly become deforested
as transport networks, particularly roads (Figures 2, 3 and 4) expanded in length
(Arbhabhirama et al. 1988; Hirsch 1987, 1990). It is to that source that one, possible
pressure groups, such as we have had in the west for quite a time now, must look with
hope that for the future of the rainforest. When the people of countries with rainforest
realise that it is a treasure something may be done. As yet, however, there appears to be
nothing on the political scene comparable to the 'Green' parties of western Europe.

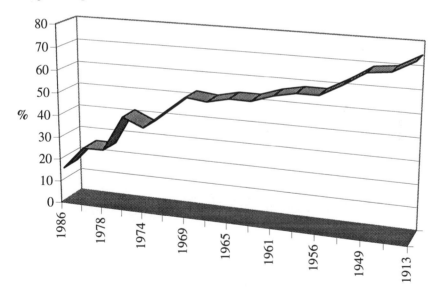

FIGURE 2.   Thailand: percentage area under forests 1913-1986 (compiled using data
tabulated in Hirsch 1987)

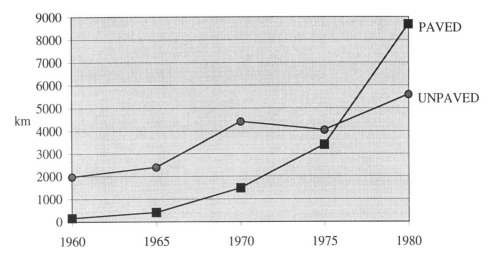

FIGURE 3. Thailand: expansion of provincial roads 1960-1980 (compiled using data tabulated in Hirsch 1987)

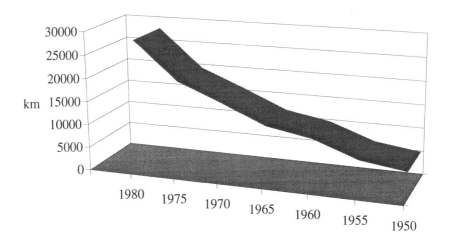

FIGURE 4. Thailand: increase in road length 1950-1980 (compiled using data tabulated in Hirsch 1987)

Protection of at least some areas of tropical rainforest gives the countries which possess them a sustainable asset of economic importance. Thailand and increasingly Indonesia and Malaysia have growing revenues from tourism, an alternative source of foreign exchange to extraction and export of tropical hardwoods to Japan, the U.S.A. and western Europe. Not all tourists want to fry beneath coconut palms on some tropical beach being waited upon hand and foot. For a long time now some of the wealthier and many of the less wealthy but fit and energetic visitors to Thailand, for instance, have enjoyed the so-called hill-tribe treks in the lower montane forest of the borderlands with Myanmar. The forest is part of the attraction, the tribal peoples of differing cultures who live there are another. Of course, there are problems. The hill tribes are shifting cultivators, populations are growing, and fallows are becoming shorter, and there is a growing demand for land from the Thai lowlanders. Additionally, the excessive number of tourists visiting easily reached villages is leading to rapid acculturation, but tourism is not having a direct impact on the fate of the forest. As in many other parts of the tropical world it is population growth which is the real problem, but this is an issue outside the immediate scope of the series of essays presented here.

Eco-tourism is one way of trying to blend conservation and development. Recently marketing of 'sustainably-produced' rainforest products has been put forward as the key to saving the forest, without consultation with the peoples of the forest, of course (Gray 1990). The exploitation of rattans which occur sporadically in the forest and are used to make cane furniture has led to a diminishing abundance (Whitmore 1980) but these could be grown.

Reafforestation and conservation of the flora and fauna are mutually antagonistic although reafforestation can, at least, reduce the rate of soil erosion. However, as Thomas indicates later in this volume, even soil erosion is not always entirely negative, as material from one area may be beneficially deposited in another. Nevertheless, the danger of reafforestation of common lands forcing displaced peoples to clear the remaining natural forests is an issue of growing concern (cf. Lohmann 1990).

It is sometimes forgotten that human impact on the remaining tropical forests is not confined to clearing of the forest itself. There are, it is true, a large number of publications concerned with the affect of mining operations on comparatively small areas but there is less which directly concerns itself with impact of flooding of land and general alteration of natural hydrological processes associated with dam construction and resettlement of the displaced populations on tropical rain forest. The literature on the Thai Nam Choan dam project is one exception (Cox 1987; Hirsch 1987).

It should be noted though (Reading et al. 1995) that there are at least 63 countries and large islands with territories within the humid tropics and the countries concerned have around 405 of the world's population, over 14 billion people. Almost all rely on agriculture to support their local and national economies, and this is where land hunger becomes so important to the future of the various types of tropical rainforest. It is also what makes sustainable development such a major concern.

4. Tropical rainforest in the colonial period

The early thrust of British, Portuguese, and Dutch colonialism was in the tropical rainforest of Southeast Asia, reliance upon spices (nutmeg, cloves and pepper) obtained through trade was replaced by small-scale planting.   In contrast, the Spanish were concerned with conquest, treasure hunting and spreading the Roman Catholic religion in the Americas and then the Philippines, competing with the Portuguese whose main long term impact has been in Brazil.   The French entered the scene later and the Belgians (the Congo), Germans (East Africa, New Guinea, South West Africa) and Danes (the Danish West Indies) later still, but it was the French and the Belgians, particularly the French, who had the most lasting influence, especially in the carve up of Africa.   This period of colonial consolidation and competition between the various European countries was as much to do with politics as the attempt to exploit tropical raw products which could not be grown in Europe and, particularly in Africa it left boundaries between states which were often vaguely defined and divided ethnic groups, a source of turmoil to the present day.   The U.S.A. came late on the scene sequestering Spanish territories in the Caribbean and the Philippines in Southeast Asia.

Throughout the rainforest world in the 19th century small-scale clearing to plant was replaced by larger scale, almost factory farming, of crops desired in Europe: tobacco, cocoa, coffee, rubber, oil palm, etc.   Useful plants were introduced from one part of the tropics to another and botanic gardens were established as centres for plant introduction and trial at places such as Singapore, Bogor in west Java, and Rio de Janeiro (Whitmore 1990).   With the cultivars, and with the coastal trading ports, came the   weeds of agriculture and waste places.   Wherever there was settlement, or on the fringes of settlements, sometimes aggressive alien species arrived and spread, transforming the flora of regrowth areas.

The early perception of the tropical rainforest was of luxuriant vegetation growing on what must be very fertile soils.   The truth of the matter is that, while not all soils which occur in the tropics are as nutritionally poor for plant growth as some literature seems to insist (soils derived from limestone and basalt are generally rich in the exchangeable bases) a substantial number, particularly in the old shield areas of Africa and South America are. In the tropics, depending again on parent material, soils derived form younger rocks or transported sediment are likely to be chemically richer than those from older rocks but many tropical soils suffer from a phosphate fixation problem.   There may be sufficient phosphate in the soil for plants to thrive but it is generally held in a form which makes it unavailable to those plants.

The soils of the tropical rainforest area proper have often been formed over millions of years, are frequently deep, well weathered, intensively leached heavy clays which are difficult to work and the clay fraction is dominated by kaolinite which has a lower cation exchange capacity than soils derived from basalt, e.g. those of the Deccan Plateau of India, which are also heavy but are dominated by 2:1 lattice clay minerals such as montmorillonite, which have a greater surface area to which the exchangeable bases and water can adhere.

Soils of the monsoon forest areas tend to be slightly chemically richer as they are less well leached (Young 1976) and may have some weatherable material left but clay is commonly moved down profile and may form an impermeable textural B horizon.

The soils of the montane forest areas, so often neglected in the academic and agronomic literature, in contrast, tend to be similar to those of temperate areas (brown earths, often acidic, and various kinds of podzol), although the climatic component of the soil-forming factors differs as both temperature and precipitation receipt are usually higher. They are shallower than those of the lowlands and are less intensely leached, so they are chemically more fertile.

Mangrove soils are peculiar. These so-called acid sulphate soils, which are not confined to the tropics but are also present in temperate estuarine areas (the Dutch call them 'cat-clays' because they smell like cats urine because of their high sulphur content) are usually clay rich and become highly acidic if they are drained rapidly (Dreisen 1974). The sulphates are transformed to sulphides and sulphuric acid and water are released in the process.

Many attempts at reclamation have rendered land barren. The watertable needs to be lowered slowly. Then the rainfall leaches the acids down below plant rooting depth in the profile but these soils also commonly have a high aluminium content and more than a few parts per million of aluminium in a plant available form is toxic to most plants. Rice is one of the more resistant plants. The iron content is also often particularly high but this is less of a problem. Cultivated plants require iron in trace amounts for proper growth anyway. Indonesia has large areas of acid sulphate soils particularly in Sumatra and Kalimantan and in recent years there has been a tendency to use mangrove areas for various types of fisheries activities rather than clear them for wet rice cultivation.

It is now well known (Longman and Jenik 1987) that the tropical rainforest has an almost closed nutrient cycling system and that once the forest is felled this is broken down and the chemical elements contained within the vegetation itself is lost and the plants replacing it add less organic detritus to the surface to be mineralized and humified by the decomposing micro-organisms. So, the richness of the soil is apparent rather than real, most tropical soils, the product of a fragile geomorphic environment, are, not surprisingly fragile too.

5.   The structure of this book

The structure of this book aims at following a natural order. In Chapter 2 it looks at the geomorphological background in relation to what Thomas (this volume) calls landscape sensitivity, in a temporal and spatial context first, stressing the importance of long term change, which extends back far beyond the time of first human occupation, but which had, and still does have, impact today on the usage of rainforest areas, and the often devastating effects of the sensitivity to extreme precipitation events which relates to the kind of materials that comprise the tropical landscape today and which is most readily traced for recent times. Soil evolution and geomorphic evolution are intimately related. It is, therefore, logical that soil sensitivity should be considered next and Nortcliff (Chapter 3) develops the theme introduced above that not all tropical soils are chemically poor, and explores soil-tree interrelationships, and the role of rainforest in soil conservation, but suggests that rainforests are not economically valuable assets within the context of modern societies; that they have been cleared to release the soil for other uses which are viewed as of economic importance. In Chapter 4 the editor tries to draw together the strands of the long term usage if rainforests by people, initially as the home of foraging societies, and gradually as a source of valuable plants, then as an area to be cultivated or grazed, drawing mainly upon the evidence from archaeology and palaeoecology. The topic of technological innovations in tropical agriculture, a theme fairly neglected, with some notable exceptions, e.g. Glover and Higham (1996), since the activities of cultural geographers like Sauer (1952) and his ethnographer predecessors within the Germanic academic tradition (Isaac 1970). Furley (Chapter 5) carries some of these approaches forward within a Central American context while Ellen (Chapter 6) explores the relationships between the indigenous perception of the forest, extraction and conservation and Fraser (Chapter 7) looks as the social, economic and land-use planning aspects of forest clearance in Indonesia from a foresters perspective, tracing the history of commercial logging from its origins in Dutch colonial times in Java during the 19th century through to developments in recent land-use planning and relates forest removal to demographic trends and political considerations. Towards the end of his chapter he investigates reafforestation programmes which he states usually results in monotonous plantation monocultures. Monoculture is very often of *Eucalyptus* and Barrow (Chapter 8) considers the economic advantages and biodiversity disadvantages of this. This brings us to a consideration of the future, specifically the climatic impacts of future rainforest destruction on local, regional, and world climates as modelled by McGuffie *et al.* (Chapter 9) before the editor attempts to draw the various strands of the arguments presented in each chapter together within the terms of the general theme of the book: human activity and the tropical rainforest, the past, present and possible future.

References

Anh, Vu Tuan (1994). *Development in Vietnam: Policy Reforms and Economic Growth*, ISEAS, Institute of Southeast Asian Studies, Singapore.

Anon. (1989). The transmigration fiasco, *Geographical Magazine*, **61** (5), 26-30.

Arbhabhirama, A., Phantumvanit, D., Elkington, J. and Ingkasuwan, P. (1988). *Thailand Natural Resources Profile*, Oxford University Press, Oxford.

Babcock, T.G. and Cummings, F.H. (1985). Land settlement in Sulawesi, Indonesia. *Malaysian Journal of Tropical Geography*, **10**, 12-25

Brunig, E.F. (1971). On the ecological significance of drought in the equatorial wet evergreen (rain) forest of Sarawak (Borneo), in J.R. Flenley (ed.), *The Water Relations of Malesian Forests*, University of Hull, Department of Geography, Miscellaneous Series No.11, Hull, pp. 66-88.

Budiardjo, C. (1986). The politics of transmigration, *The Ecologist*, **16** (2/3), 111-117.

Conway, G.R. and McCauley, D.S. (1983). Intensifying tropical agriculture: the Indonesian experience, *Nature*, **302**, 288-289.

Cox, B.S. (1987). Thailand's Nam Choan Dam: a disaster in the making, *The Ecologist* **17** (6), 212-219.

Devan, J. (1994). *Southeast Asia: Challenges of the 21st Century*, ISEAS, Institute of Southeast Asian Studies, Singapore.

Glover, I.C. and Higham, C.F.W. (1996). New evidence for early rice cultivation in south, southeast and east Asia, in Harris, D.R. (ed.) *The origins and spread of agricultuure and pastoralism in Eurasia*, UCL Press, London, pp.413-441.

Gray, A. (1990). Indigenous peoples and the marketing of the rainforest, *The Ecologist*, **20** (6), 223-227.

Grubb, P.J. (1971). Interpretation of the 'Massenerhebung' effect on tropical mountains, *Nature*, **229**, 44-45.

Grubb, P.J. (1974). Factors controlling the distribution of forest-types on tropical mountians: new facts and a new perspective, in J.R. Flenley (ed.), *Altitudinal Zonation in Malesia*. University of Hull, Department of Geography, Miscellaneous Series No. 16, Hull, pp. 13-45.

Harrison, P. (1977). Indonesia: food, population, land. Can Indonesia farm the swamps? *New Scientist* 22/29 September, 804-805.

Hirsch, P. (1987). Nam Choan: benefits for whom? *The Ecologist*, **17** (6), 220-222.

Hirsch, P. (1990). *Rainforest Politics: Ecological Destruction in Southeast Asia*, Zed Books, London.

Jacobs, M. (1987). *The Tropical Rain Forest*, Springer, Berlin.

Isaac, E.. (1970). *Geography of Domestication*, Prentice-Hall, Eaglewood Cliff, New Jersey.

Isamangun and Dreissen, P.M. (1974). The acid sulphate soils of Indonesia, in *Agricultural Cooperation Indonesia-The Netherlands Research Reports 1968-1974*, Section II: Technical Considerations, Republic of Indonesia, Ministry of Agriculturw, Jakarta, pp. 206-218.

Koppen, W. (1936). Das geographische System der Klimate, in *Handbuch der Klimatologie*, vol. 1. Borntager, Berlin.

Holdridge, L.C. (1967). *Life Zone Ecology*, Tropical Science Centre, San Jose, Costa Rica.

Lohmann, L. (1990). Commercial tree plantations in Thailand: deforestation by any other name, *The Ecologist*, **20** (1), 9-17.

Longman, K.A. and Jenik, J. (1987). *Tropical Forest and its Environment*, 2nd ed., Longman, London.

Otten, M. (1986). 'Transmigrasi': from poverty to bare subsistence, *The Ecologist*, **16** (2/3): 71-76.

Reading, A.J., Thompson, R.D. and Millington, A.C. (eds.) (1995). *Human Tropical Environments*, Blackwell, Oxford.

Savage, J.M., Goldman, D.P., Janos, D.P., Lugo, A.E., Raven, P.H., Sanchez, P.A. and Wilkes, H.G. (1982). *Ecological Aspects of Development in the Humid Tropics*, National Academy Press, Washington, DC.

Secrett, C. (1986). The environmental impact of transmigration, *The Ecologist*, **16** (2/3), 77-88.

Steenis, C.G.G.J. van (1962). The mountain flora of the Malaysian tropics, *Endeavour*, **21** (82-83), 183-193.

Sumintardja, D. (1976). Landless Javanese migrate to Sumatra, *Geographical Journal*, **48** (8), 467-468.

Takaya, Y. (1987). *Agricultural Development of a Tropical Delta: A Study of the Chao Phraya Delta.* Monographs of the Center for Southeast Asian Studies, Kyoto University, English Language Series No.17. University of Hawaii Press, Honolulu.

Thompson, G.B. (1994). Wood charcoals from tropical sites: a contribution to methodology and interpretation, in J. Hather (ed.), *Tropical Archaeobotany: Applications and New Developments*, Routledge, London, pp. 9-33.

Tricart, J. (1972). *Landforms of the Humid Tropics, Forests and Savannas*, Longman, London.

Troll, C. (1959). *Die tropischen Gebirge.* Bonner Geographische Abhandlingen, Heft 25, Dummter Verlag, Bonn.

Walker, D. and Pittelkow, Y. (1981). Some applications of the independent treatment of taxa in pollen analysis. *Journal of Biogeography*, **8**, 37-51.

Walter, H. (1979). *Vegetation of the Earth and Ecological Systems of the Geobiosphere*, 2nd ed., transl.J. Weiser, Springer, New York.

Whitmore, T.C. (1980). Potentially economic species of Southeast Asian forests, *BioIndonesia*, **7**, 65-74.

Whitmore, T. (1990). *An Introduction to Tropical Rain Forests*, Clarendon Press, Oxford.

Young, A. (1976). *Tropical Soils and Soil Survey*, Cambridge University Press, Cambridge.

Dr. Bernard K. Maloney, Palaeoecology Centre, The Queen's University, Belfast BT9 6AX, Northern Ireland

2. LANDSCAPE SENSITIVITY IN THE HUMID TROPICS - A GEOMORPHOLOGICAL APPRAISAL

Michael F Thomas

1. Introduction

The concept of *landscape sensitivity* was introduced into thinking about geomorphological systems by Brunsden and Thornes (1979) and has been developed in the context of environmental change (Thomas and Allison 1993). When resistance to change is exceeded by the magnitude of the disturbing forces the environmental system will adjust to create a new equilibrium. However, that equilibrium will not be attained if the frequency of high magnitude events, sufficient to cause further disturbance, is greater than the 'relaxation time' of the system. Some attributes of environmental systems, and equally, some areas of natural landscapes, are more sensitive to change than others. Both the materials (rocks, saprolites, sediments, soils) of the landscape and its morphological components and their arrangement (aspects of slope and landscape pattern) contribute to its sensitivity and when these, mainly landform, characteristics prove stable those parts of the landscape can be said to have a high 'factor of safety', a concept borrowed from engineering (Brunsden 1980, 1993; Brunsden and Thornes 1979). Where the landscape has very low resistance to change (high sensitivity) it may said to be 'fragile'.

These ideas have been applied far more to what might be called *temporal sensitivity*, in the context of the magnitude and frequency of formative events and the prediction of conditions under which rates of erosion and sedimentation may be increased, than to the question of *spatial sensitivity*, or the location of sources of sediment, where soil and slope instability are most likely to be induced by changes in climate and plant cover. *Sensitivity analysis* of course has other connotations (cf. Downs and Gregory 1993) and, ideally should be capable of mathematical expression, but its application to landscapes remains largely conceptual, due to the complexities of scale and lack of data.

Questions asked about sensitivity to change raise issues concerning rates of change, not only of system parameters such as runoff and sediment yield, but also amongst their external, controlling factors such as rainfall and vegetation cover. Change due to human agency may be equally complex, when viewed over long periods of time, and over large areas. It is also necessary to consider system behaviour in relation to progressive change over time and the operation of internal thresholds. When such thresholds are crossed natural systems experience what has been termed a *complex response* (Schumm and Parker 1973), and undergo internal reorganisation. This makes the interpretation of cause and effect more difficult.

2. Landscape sensitivity in the forested tropics

An illustration of the interaction of different factors may be helpful initially. Conservation practice generally emphasises the protection of steep slopes by retaining the natural forest cover, and this is usually wise, yet much of the most severe soil erosion has taken place on gentle slopes underlain by unconsolidated sandy colluvium, or has been triggered at specific, sensitive sites. Many surface materials may have had

17

B.K. Maloney (ed.), Human Activities and the Tropical Rainforest, 17-47.
© 1998 *Kluwer Academic Publishers. Printed in the Netherlands.*

their origins within climatic regimes usually drier than, and quite different from, the present, and they become subject to rapid evacuation as the system adjusts to present-day conditions.   This adjustment is normally not continuous; it is dependent on the incidence of formative (usually rainfall) events of high magnitude which may have a low frequency, with recurrence intervals of 10s or even 100s of years. However, removal of the forest cover may greatly increase sensitivity to change and lead to irreversible erosion. It is, therefore, clear that surface deposits, slope conditions, past climates and the magnitude and frequency of erosional forces are all aspects of the problem of determining landscape sensitivity. Thus this discussion is organised under a number of headings:

     i. Landscape elements and properties
    ii. Sensitivity of landscape materials
   iii. Variability of controlling climatic factors
   iv. Sensitivity of the Quaternary deposits
    v. Models describing landscape sensitivity

2.1. LANDSCAPE ELEMENTS AND PROPERTIES

Views which suggest that humid tropical landforms differ little from their temperate counterparts emphasise that both are fluvially carved landscapes subject to chemical weathering and the laws of water flow. This is correct but unhelpful; in the same way that it would be specious to argue that all vegetation responds to the DNA code and grows by a common process involving photosynthesis. This is, firstly, because the weathered mantle is thicker, more pervasive and has distinctive properties in humid tropical areas (Deere and Patton 1971; Geological Society 1990; Thomas 1994), and, secondly, because the pathways of water flow and their effects on the erosional system are strongly influenced by the weathered layer and also by distinctive soil structures. Additionally, the importance of mass movement in these environments has been seriously underestimated in the past and, where landslides have been considered important, they have commonly been attributed more to earthquakes than to climate (Garwood *et al.* 1979).

2.1.1.   *Slopes and landscape sensitivity*
Hillslope instability is to some extent dependent on slope angle. Characteristically, slopes fail at inclinations above 26-28° in Tanzania (Temple and Rapp 1972); Japan (Iida and Okunishi 1983); Sierra Leone (Thomas 1983), and also in northwestern USA (Ellen *et al.* 1988). Frequency of failure tends to reach a maximum above 35°. Jibson (1987) found a concentration of debris flows on slopes of 38-50° in Puerto Rico, and Iida and Okunishi (1983) discovered a similar relationship in Japan (max 35-40°). Slopes above 50° become *weathering limited* as gravitational forces act to transfer regolith downslope before it can attain sufficient thickness to give rise to discrete failures. However, slopes of much lower inclination (<20°) can fail during very extreme events and when denuded of vegetation.

2.1.2.   *Regolith thickness and the weathered mantle*
It is possibly more important to recognise that landscape elements develop on particular materials and at specific positions on a hillslope, with *in situ* weathered materials, or *saprolites*, preferentially preserved on less steep slopes and where the landsurface has been relatively stable for a long time.   Over susceptible rocks in the humid tropics,

weathering depths commonly exceed 30m and can reach thicknesses exceeding 100m. This means that entire landscapes of hills and valleys may be dominated by thick saprolites and derived colluvium. These often appear as multi-convex patterns of hills and intervening valley flats (often called *meias laranjas* or *demi oranges*). Such landscapes are strikingly demonstrated where new road cuttings have been excavated with great ease. In terrains of higher internal relief, thick saprolites may be confined to plateau areas or found beneath valley floor sediments. Evidence indicates that saprolite thickness tends to increase beneath interfluves in humid forest climates (Thomas 1994).

Weathering profiles exhibit characteristic zones, each of which shows different degrees of rock alteration and proportions of fresh rock (Figure 1), while some display advanced leaching or accumulation of Fe (and other minerals in smaller amounts). Where these profiles are dissected by contemporary valleys, hillslopes and soils are formed on different zones of the weathering profile, and this situation will produce important soil variations as well as variable sensitivity to erosion across the hillside.

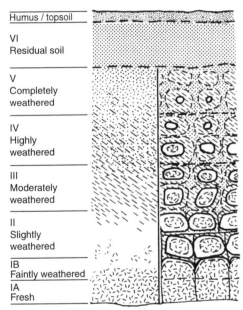

Humus / topsoil	
VI Residual soil	All rock material converted to soil : mass structure and material fabric destroyed. Significant change in volume
V Completely weathered	All rock material decomposed and/or disintegrated to soil. Original mass structure still largely intact
IV Highly weathered	More than 50% of rock material decomposed and/or disintegrated to soil. Fresh/discoloured rock present as discontinuous framework or corestones
III Moderately weathered	Less than 50% of rock material decomposed and/or disintegrated to soil. Fresh/discoloured rock present as continuous framework or corestones
II Slightly weathered	Discolouration indicates weathering of rock material and discontinuity surfaces. All rock material may be discoloured by weathering and may be weaker than in its fresh condition
IB Faintly weathered	Discolouration on major discontinuity surfaces
IA Fresh	No visible sign of rock material weathering

Idealised weathering profiles - without corestones (left) and with corestones (right)

Rock decomposed to soil
Weathered / disintegrated rock
Rock discoloured by weathering
Fresh rock

FIGURE 1. Idealised weathering profiles, shown with and without residual rock cores, according to engineering classification (modified from Geological Society, 1990 with permission). Total depth may vary from a few m to more than 100m and the thickness of individual zones from less than 1m to 10s of m.

2.1.3. Landslide debris and colluvium

Transported materials are often present in the upper zones of the regolith and can be broadly sub-divided into talus, landslide debris, colluvium and alluvium. The last three have greatest importance here and often occur together. Landslides in the forested tropics are spatially frequent, and while many leave erosional scars and expose bare rock, others may move slowly within the deep saprolite, allowing the forest canopy to adjust and survive. They often surround steep hills or take place along valley sides as in the Freetown Peninsula of Sierra Leone. Here, slide scars were mapped from large scale (1:12,500) aerial photographs and checked in the field, leading to the conclusion that the mapping was a correct representation of ground conditions (Figure 2). These slides took place within saprolites formed over basic igneous rocks and varied (Thomas and Thorp 1992) from deep seated rotational slides (slumps) to shallow translational slides (debris slides and flows). Aerial and ground reconnaissance over parts of Kalimantan also indicates the importance of landslides around most residual granite hills, while records from the Atlantic margins of Brazil confirm the importance of landsliding in that environment, affecting saprolites on granitoid and gneissic rocks (Jones 1973; Bigarella and Becker 1975; Ab'Saber 1988; Smith and de Sanchez 1992). Further examples are cited from Dominica in Reading *et al.* (1995).

Most landslide events also involve flows of liquified mud and other debris. This becomes deposited on lower slopes, and may grade into true alluvium, where the water volume is sufficient to suspend the fine clay and silt particles in the flow. Colluvial transfer of saprolite also takes place in a diffuse manner by repeated fluxes of slope (sheet) wash. However, whether this can occur beneath a forest canopy is open to considerable doubt, although runoff can be important under forest (Leigh 1982). Examples of stratified colluvium in the rainforest environments therefore attract attention as possible indicators of former conditions of drier climate and open vegetation cover.

It has been estimated that 50% of the landsurface in Sao Paulo State is underlain by colluvium (Ferreira and Monteiro 1985), and it is undoubtedly widespread throughout the rainforest zone. The complexity of colluvial/alluvial stratigraphy has been carefully documented from the Bananal area of southeastern Brazil (De Moura *et al.* 1989). However, difficulties are encountered in the recognition and definition of colluvial sediments and their formative processes. Some homogenous and possibly mixed upper zones of sandy podzolic weathering profiles have been considered the products of colluviation (Heyligers 1963) but may be due to extreme leaching of unconsolidated sediments (Righi and Chauvel 1987; Thomas 1994).

On most humid tropical slopes a combination of current and past processes has led to complex variations in the substrate, and these influence water flow and soil development, often in decisive ways. This *anisotropy* in the rainforest environment has received scant treatment by soil scientists and ecologists, but may be expected to influence the pattern of landscape sensitivity and therefore the age and structure of the vegetation.

2.1.4. Floodplains and alluvial deposits

While alluvial deposits are obviously characteristic of river floodplains, mixtures of colluvium and alluvium form gully fills, and may be intercalated on lower hillslopes, marking former sites of incision and sedimentation in the landscape. These are usually

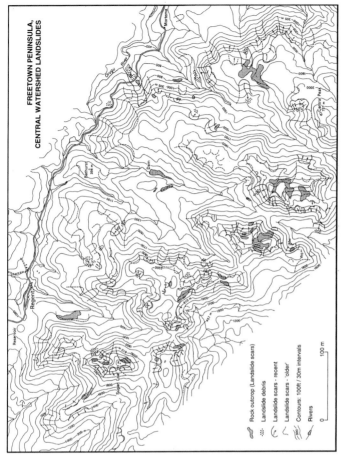

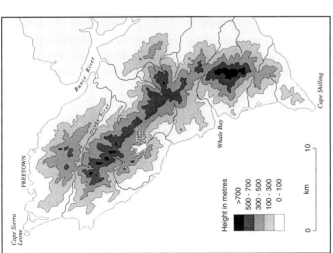

FIGURE 2.   Landslide scars flanking the main ridge of the Freetown Peninsula, Sierra Leone (mapped from 1:12,500 scale air photographs onto the 1:10,000 scale maps). Most slopes >26° exhibit landslide scars or debris (own work).

only seen during excavations or where forest has been cleared, but inspection of many road cuttings and some mining operations in west Africa has convinced the author that old *palaeochannels* filled with sediments often form networks under the forest.

Because of the links between catchment and floodplain it is obvious that alluvial sediments reflect overall catchment conditions. Thus, the widespread deep chemical alteration of rocks outside of mountainous regions reduces the availability of coarse bedload and increases the proportion of sand reaching the river. Bank erosion and landslides reaching the channel can also inject quantities of fine suspended sediment, much of it clay, into the stream. Tropical floodplains are, therefore, frequently dominated by fine grained sediments: mainly sandy, channel sediments and thick, clay-rich overbank deposits. Exceptions, however, are common, because coarse cobbles and even boulders can form the stream bed in steep and rocky terrain, as in other parts of the world.

Floodplains evolve over time and most rivers store sediments for periods of 10^4 years, sufficiently long to record the major climate changes at the last glacial maximum (LGM) around 18,000 B.P. Many parts of the humid tropics were significantly drier for several millenia during this period, but became rapidly wetter during the climate warming at the end of the Pleistocene (Thomas and Thorp 1992, Thomas 1994). These events are recorded in the floodplain stratigraphy of tropical rivers (Thomas and Thorp 1980, 1992; van der Hammen 1991; van der Hammen *et al.* 1992a, b), which frequently have buried channels containing coarse sediments dating from periods of increased flood discharges at 12,700-8000 B.P.

At this time many rivers experienced channel cutting and rapid channel aggradation, but subsequently began to stabilise as the forest vegetation recolonised slopes that may have been under savanna or deciduous woodland for several millennia at the glacial maximum. This in turn led to the fining of the sediments and a lessening of the total sediment load. Fine grained overbank sediments accumulated by vertical accretion and lateral channel shifts became inhibited by the cohesiveness of the sediment and due to the protectiveness of the bank vegetation. Often such river channels appear stable, but can experience sudden *avulsion* during major floods, when the river spills over the natural levees and cuts a new and shortened course downstream. Surprisingly, Downs and Gregory (1993) state that rather few studies of river channel sensitivity have been made. But, because they are the source of flooding for many populous areas, the subject is of some importance.

Many small tributary valleys, especially on planate landscapes have largely channelless, flat floors and are subject to frequent or seasonal flooding. Although particularly common in the African savannas, where they are known as *dambos*, or *vleis*, humid tropical forms such as the *baixas* of Brazil and the *bolis* of Sierra Leone are also found, where they are known as inland valley swamps (Figure 3). These are one type of a number of wetland systems that are particularly sensitive to hydrological changes consequent upon drainage, or channlisation of overland flow. Such valleys are often infilled with sediment, possibly dating from a period of drier climate, and these deposits are suscptibe to rapid erosion when delicate equilibria are disturbed. In their undisturbed state the may form the hydromorphic soil environments discussed below.

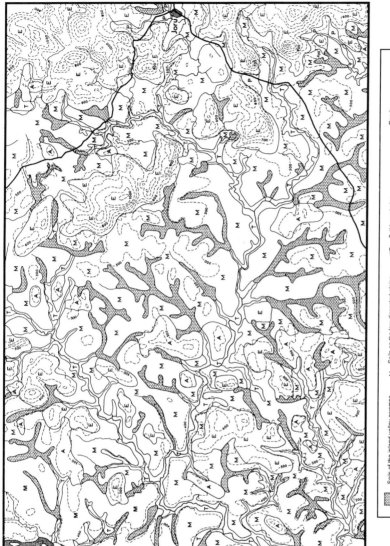

FIGURE 3. Soil-landform map of part of Sierra Leone, showing extent of inland valley swamps (*bolis*) (adapted from FAO, 1968).

2.2. SENSITIVITY OF LANDSCAPE MATERIALS

These highly variable substrates represent an equally variable sensitivity to environmental changes and re-mobilisation in the erosional system. An approach to closer definition of these properties should begin with a consideration of the residual weathered material, or saprolite, itself.

2.2.1. *Saprolite properties and landsliding*

Details of saprolite properties are to be found elsewhere (Geological Society 1990; Thomas 1994; Ollier and Pain 1996), and they are determined largely by the degree and type of rock alteration. This covers the amount of mineralogical change (degree of weathering) and the nature of secondary and residual minerals, especially clays and Fe and Al sesquioxides (type of alteration). These chemical and mineral characteristics in turn affect the rock mass properties of the saprolite. The following factors are particularly significant:

  i. clay mineralogy, which determines the presence of expansive or dispersive soils; the presence of Al and Fe oxides and the resultant percentage of clay-sized particles

  ii. micro-aggregation of (mainly kaolinitic) clays, and voids ratio as determinants of permeability (infiltration capacity) and runoff coefficient

  iii. profile discontinuities due to transported and illuviated materials (Fe, clay), leading to lateral water flows

  iv. 3-D, spatial anisotropy due to residual rock fabric discontinuities and formation of macropores (pipes)

  v. profile depth and the nature of the interface with unaltered rock (weathering front).

Most saprolites in the humid tropics evolve towards highly leached, ferrallitic (Duchaufour 1982) materials dominated by kaolinite (or halloysite) and containing significant amounts of Al and Fe sesquioxides (gibbsite, goethite). This mineralogy allows the saprolite to retain considerable shear strength even when it contains large quantities of water (Reading *et al.* 1995). Further incongruent breakdown of the kaolinite can lead to more specifically allitic (or bauxitic) residues, while congruent kaolinite dissolution must lead to accumulation of quartz sand. Where this occurs in swamp environments, or where organic acids act as chelates to remove Fe and Al in organo-metallic complexes, hydromorphic podzols are formed (Duchaufour 1982; Righi and Chauvel 1987; Lucas *et al..* 1987, 1988).

Except over very quartz-rich rocks these forest soils develop a high clay content, typically 40+/-20%, but retain a high infiltration capacity due to the formation of micro-aggregates involving the amorphous Fe and Al sesquioxides. Figures for *infiltration capacity* vary, but commonly exceed 100 mm h^{-1}, and infiltration rates during the first 30 minutes of a storm have been measured at more than 280 mm h^{-1} near Ibadan in Nigeria (Nye 1954, 1955). Here rainfalls of 100 mm h^{-1} intensity (over 30 min) occur on average every 2.8 years, but those exceeding 200 mm y^{-1} (over 7.5 min) are rare, with a return period of 100 years (Lal *et al*. 1981; Lal 1986). Peak rainfall intensities are attenuated by canopy interception under rainforest, and although the terminal velocity of large leaf drips and the volume of stem flow have erosional potential, the ground surface is protected by roots and plant litter, and lateral transfers of surface soil remain very limited.

The high figures for infiltration capacity also reduce surface runoff in many forest areas to minimal amounts throughout most years, even on slopes of 20°. However, exceptional storms can lead to significant surface flows under two conditions: firstly, during storms of extreme intensity, such as occur in hurricane/typhoon zones; secondly, after prolonged rainfalls as all the available pores become filled so rapidly that deep percolation or lateral throughflow cannot take place. On steep slopes strips of forest are torn away by these events, but the regolith properties often ensure that channel flow is not continued long after the storm, while the scars heal over with new tree growth. However, the reactivation of such slopes can be frequent, as seen in Puerto Rico (Jibson 1989; Larsen and Torres-Sanchez 1996). It is a common experience in hilly areas of rainforest to find very few large trees, which suggests a replacement cycle of 10^1 y rather than 10^2 years.

Conditions are seldom uniform across a slope, and seepage leading to runoff is likely to start first on lower slopes and in *hillslope hollows*, many of which lead into the unchannelled valleys that pattern most hillslopes. Such hollows have been the subject of research in humid temperate areas (Pacific NW, USA), where they are found to contain colluvial wedges of transported sediment (Dietrich and Dorn 1984; Dietrich *et al*. 1986). These are highly sensitive sites which are subject to saturated overland flows of water (Reneau and Dietrich 1987) and to periodic, shallow debris flows (Reneau *et al*. 1986). During extreme rainfall events such small valleys can experience rapid excavation and stripping of the colluvium, followed by the formation of shallow channels. It also appears likely that this kind of sensitivity is influenced by threshold values for the accumulation of colluvium in the hollows: sensitivity to events which can lead to rapid removal of the debris downslope as the depth of sedimentation and slope of the ground surface increases.

Landslides in saprolite can also take the form of larger *translational slides* which often strip the weathering front to expose rock slabs, and deep seated *rotational slides* or slumps that occur where there is a thick saprolite or layered sedimentary formations containing an incompetent bed. Translational slides often strip 5-10m of regolith from steep slopes of 28-38°, and they appear to be a major cause of rock exposures in areas of igneous and metamorphic rocks. They are favoured by the formation of stress release joints in rocks along steep valley walls. These open fractures act as subsurface conduits for infiltrating water and encourage rock weathering. The weathered layer becomes sensitive to landsliding partly because it inherits these discontinuities which weaken the weathered mass covering the slope.

Rotational slides possess curved shear planes that extend deep below the surface, and in weathered terrain these failure planes develop within the saprolite. These slides are less likely to be triggered by single storm events, although they may induce renewed activity. Most commonly they arise where basal support for the slope has been removed, as when a river undercuts a steep slope, or when artificial excavations are undertaken for road and dam construction or other purposes.

These and other types of landslide are much more frequent in tectonically active zones containing very high relief, severely deformed sedimentary rock sequences, and fault scarps. Here, climatic factors combined with high relief and, often, with repeated seismic shocks increase landslide activity (Simonett 1967; Garwood et al. 1979). It is not always recognised that there is such an interaction between the weathering systems and earthquake impact. Deep karstification, with the formation of underground caverns weakens the rock fabric and can predispose large rock masses to fail under seismic shock waves (King et al. 1989), while deep weathering of silicate rocks can also contribute to the severity of landslide activity during earthquakes.

Two apparently contradictory views of landslide activity can be advanced. Firstly, it has been observed that fresh landslides often occur on the same sites as palaeoslides (Nilsen and Turner 1975). This usually results from the reactivation of old shear surfaces and can occur during construction, where the basal support is removed from the slide debris. However, the propensity for some sites to fail repeatedly may also reflect the geology and slope conditions. Secondly, where slides occur in a relatively shallow regolith (<10m), repetition of these events on the same sites must depend on the renewal of the regolith layer. This is sometimes referred to as the 'ripening' of the slope. To re-form an in situ weathered layer 5-10 m thick could take 10^3-10^4 years (at a weathering rate of 0.25 mm y^{-1}), but of course slow soil creep can mantle the hillslope more quickly by drawing on the saprolite store from a higher level on the slope.

In general, it is possible to say that deep seated slides often have long formative histories (10^2 y) and become activated repeatedly during periods of high rainfall, while some shallow slides, such as debris avalanches and flows appear to have short life spans and may heal over after a single event lasting some hours. These shallow slides can, however, be regarded as very active on steep hillslopes in climates prone to high magnitude rainfall events. They probably constitute the dominant slope process in forested terrains within hurricane belts such as the Caribbean and they have been widely studied in Puerto Rico (Jibson 1989; Larsen and Torres-Sanchez 1996). A major storm which laid waste to a small area of 180 km^2 in the Serra des Araras, 50 km west of Rio de Janeiro in Brazil in January 1967 was estimated to have led to >10 000 slope failures within this small area. Most of these were debris flows and many occurred beneath undisturbed forest (Jones 1973).

2.2.2.   Properties of colluvium and alluvium
Sediments reaching the valley floors in rainforest areas have commonly begun as weathered layers and have been transported in stages over periods of 10^1 - 10^3 years, possibly having accumulated on lower valley slopes as colluvium for prolonged periods. In most cases the resultant sediment is in an advanced stage of weathering, containing very few rock pieces, and is comprised mainly of quartz sand and clays in variable proportions according to the original rock composition.

The properties of the colluvium depart significantly from those of the saprolite formed in site, and, in particular their infiltration capacities may be low; less than 20 mm h^{-1} was measured in Kenya (Singh et al. 1984). Much higher figures can, however, occur on very sandy materials. The colluvial transport appears able to break down soil aggregates and natural bonds between clays and oxides. It will also lead to sorting of sediments by running water, so that sandy and clayey layers may be intercalated. Subsequently clay eluviation from upper layers and illuviation at depths of perhaps 1-2 m can produce sandy topsoils which are sensitive to rainsplash and lateral transport, and also clay pans which may lead to lateral throughflow with excavation of tunnels (pipes).

The fine grained alluvium of many tropical floodplains is also readily leached by infiltrating rainfall on the almost flat slopes and some specialised soils, partly discussed before, can evolve. However, the general tendency for stratification of alluvium frequently leads to lateral throughflow along discontinuities, and the discontinuities may develop into open conduits or *soil pipes*. These often develop into gullies that enlarge by headwall retreat from the river bank, or by collapse under abnormal loading (e.g. by vehicles and some large mammals). The presence of pipes in sedimentary formations is widely reported and observed in the humid tropics (Loffler 1974, 1977; Thomas 1994).

The zones most sensitive to erosion within rainforest landscapes appear therefore to include at least the following:

i. **steep hillslopes**, especially those between 28-45°

ii. **hillslope hollows** with very variable slopes that act as collecting areas for convergent groundwater seepage

iii **zones of stratified colluvium**, which are generally on lower slopes

iv.**floodplain sediments or channel infills** containing intercalated sands and clays.

A descriptive model of these conditions is shown in Figure 4.

2.2.3.   *Hydromorphic environments and biochemical sensitivity*
Many fluvial and some coastal environments contain areas of so-called *white sands*. These materials are generally hydromorphic tropical podzols (Brunig 1974; Duchafour 1982), and are often developed in coastal sands, most of which prove to be alluvial rather than marine in origin. Because these sediments are characteristic of some distinctive tropical forest environments, containing species poor 'heath' forests such as the *kerangas* of Sarawak and Kalimantan on the island of Borneo (Brunig 1974), they merit separate attention.

In the inner humid tropics, with their high rainfalls and short dry periods, ferrallitic soils have a pronounced tendency to evolve into podzols (Chauvel *et al.* 1987; Leucas *et al.* 1987, 1988). Many soils over igneous rocks are ochreous in their upper horizons and show pronounced leaching and clay eluviation. Sediment fluxes from slopes therefore contain much sandy, iron depleted, material, and when this arrives into the (hydromorphic) valley-floor environment, it becomes subject to strong lateral flows of water containing significant amounts of organic acids. This leads to further leaching and removal of clay. The upper horizon is also subjected to continued vertical leaching and eluviation in a second cycle of *lessivage*.

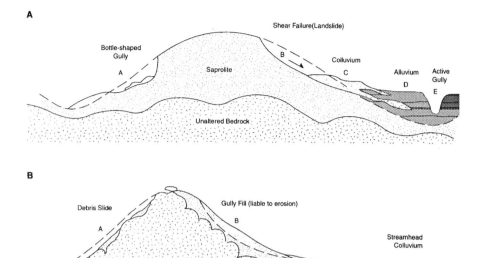

FIGURE 4. Incidence of erosional events in response to rising rainfall in the Equatorial tropics [A], but declining rainfall in marginal rainforest climates [B]. This is, of course, speculative, but increasing frequencies of high magnitude rainfalls [A] would lead to a greater incidence of landsliding in hilly terrain, while dry years within a seasonal forest area could lead to widespread fires [B] which might be followed, in wetter years, by frequent debris flows and erosion by direct runoff.

This signifies a *biogeochemical sensitivity* to progressive impoverishment of the residues of weathering and as it takes place, the forest cover clearly evolves, becoming dominated by species tolerant of such environments. It is these forests, growing in an almost sterile soil environment, that are probably not able to regenerate once the delicate nutrient cycle has been destroyed. This is in stark contrast to conditions on steep hillslopes where actively weathering rock comes close to the surface and plant nutrients are abundant. Here, forest can reclaim savagely guttered hillslopes just a few years after major storms.

A possible corollary of this observation might be that, as leached, sandy topsoils evolve on a wide range of terrains in the perhumid tropics, the slow shedding of sand from the slopes may be beneficial, allowing the soil forming process to tap into stores of less altered saprolite at depth. Occasionally, we might ask ourselves if all soil loss is detrimental to present and future land use.

2.2.4. Coastal environments

Coastal environments and their sensitivity to change have not been considered here, but it should be recognised that the fine grained sediments, mainly clays, that are removed from river catchments are flocculated in the marine environment and may become stabilised by the growth of mangroves. The sandy channel sediments also progress seawards and are swept along the shore to form beaches and barriers, behind which brackish or freshwater lagoons are created. Organic matter accumulates in these basins as well as fine sediments, and peats may be formed. The coastal complex of alluvial floodplain, peat swamp, lagoon, beach barrier and mangrove contains formations that are rather easily eroded or modified by changes in the alluvial and coastal systems.

2.3. VARIABILITY OF CONTROLLING CLIMATIC FACTORS

The study of climatic change over periods of 10^3-10^5 years has traditionally been concerned with major shifts in climate, leading to changed patterns of vegetation and soil persisting for thousands of years, and linked to glacial cycles in the northern hemisphere during the Quaternary. This approach has allowed certain types of sediment to be allocated to past climatic regimes in the tropics, and more specifically to link changed rates and styles of activity on tropical landscapes to both wetter and drier phases of climate. By contrast, present day climate has been analysed on the basis of the magnitude and frequency of events, principally storms and droughts. Time series analysis of such data has also allowed changes in the means to be detected on a decadial basis for periods ranging from 50-150 years, but very few stations have recorded meteorological data for longer than 100 years in tropical areas.

It is now more widely recognised that climatic change cannot be interpreted simply as a shift in the means of temperature or rainfall, but must encompass altered frequencies of storm events and storm sizes, linked to changes in wind patterns and other parameters. Our ability to detect these changes is increasing rapidly with the use of AMS radiocarbon dating of finely subdivided sedimentary sequences. Also, as data on temperature change increases, it has become recognised that significant climate shifts can occur over periods of a few decades and specific climatic patterns may persist for hundreds rather than thousands of years (Street-Perrot 1990; Gasse and van Campo 1994).

In studies of landscape sensitivity, therefore, it has to be recognised that patterns of events and timescales of change are different aspects of the same phenomenon of climatic variability. Yet, major climate shifts clearly have occurred in the past and have led to important changes to system inputs and behaviour. The central problem is to develop greater understanding of the sensitivity of landscapes to changes on different timescales. For convenience such a discussion can be divided rather arbitrarily:

    i. sensitivity to individual events lasting for periods of minutes, hours or days (10^{-3}-10^{-2} y)

    ii. effects of periods of prolonged wetness or drought persisting for weeks, months or years (10^{-1}-10^{0} y)

    iii. cumulative impacts of changes in the magnitude and frequency of events, particularly of rain storms and associated wind systems over decades or centuries (10^{1}-10^{2} y)

    iv. longer term effects of shifts in the means of system parameters such as temperature and rainfall persisting for centuries or millennia (10^{2}-10^{3} y).

Even with this guide problems are not avoided.   Firstly, because these categories are arbitrary sub-divisions of a continuum and they lack specific information on **rates** of change, except in the case of individual storm and flood events. Secondly, because while we lack detailed understanding of past climates and vegetation patterns, it is certain that the sequence of changes following progressive warming and increased rainfall such as occurred in the tropics at the Pleistocene-Holocene transition (12,000-9000 B.P.), will have been different from any record we have of actual rainfalls during the modern era.

2.3.1. *Sensitivity to extreme events*
Floods, major increases in sediment yield and shallow debris slides and flows all result from short-term extreme rainfall events, commonly measured over periods from 30 minutes to 24 hours. But all these processes are influenced by *antecedent conditions*, particularly in the wetness of the soil and substrate resulting from rainfall inputs over days, weeks or months.

In equatorial and some mountain areas, where heavy rainfalls occur on a daily basis, small streams draining steep hillslopes are fed by groundwater almost continuously and discharges rise rapidly during storms, even where overland flow is not evident. Many headwaters drain from hillslope hollows and, during extreme events, debris flows from the stream heads may engulf the channel downstream, gathering discharge, and more sediment by channel and bank erosion as the flood wave (mud/debris flow) tears through the forest. Data from a number of locations suggest that such events commonly result from storm sizes of 200-200 mm in 24h, containing pulses of rainfall exceeding 100 mm h^{-1}. In Puerto Rico, Larsen and Simon (1993) found that landslide triggering storms could be fitted to threshold values according to the following relationship:

$$I = 91.46 \, D^{-0.82}$$

where I is rainfall intensity (mm h-1) and D is duration in hours. Worldwide, Caine (1980) found a fit with $I = 14.42 \, D^{0.39}$, but in Puerto Rico, for storm durations less than 10 h, 3x as much rainfall is required to produce landslides as in temperate areas, however, as the duration increases, so the amounts converge. Nevertheless, intensities as low as 13.8 mm h^{-1} were found to trigger landslides on slopes over 12° (Larsen and Torres-Sanchez 1996).

Antecedent conditions are often built into landslide prediction models (Lumb 1975), and when widespread devastation takes place and involves deep seated slides, the build up of wet conditions over a period is implicated. Lumb (1975) used the previous 15-day rainfall for Hong Kong and found that landslide disasters followed c. 350 mm of rain over this period and at least 100 mm in the 24 h preceding the slides. But such figures are of empirical value only, each locality responding to different specific controls. On the Freetown Peninsula, Sierra Leone, major landsliding occurred in August 1945 after an unprecedented 1121 mm over 5 days, culminating in 401 mm on the day of the slides. The event triggered at least one major translational slide 1 km long on a 38° slope and also a deep seated rotational slide in thick saprolite. These sites were visited in 1967, 1974 and 1983 and it was clear that no major changes had taken place during this period. Indeed the slide scar produced in 1945 was probably similar in extent 40 years later. Slopes which were sensitive to change in the August 1945 rainstorm were, therefore, not sensitive, on the same or on adjacent sites during heavy rainfall in subsequent years.

The main ridge of the Freetown Peninsula is, however, facetted by traces of former landslides, and in one area of the main ridge the occurrence is 6.6 km^{-2} (plotted from 1:12,500 aerial photographs). We have no idea of the time taken to produce this number of slides, which does **not** include the traces of any superficial debris slides/flows, but is restricted to translational and rotational slides, most of which exceed 30 m in width (Figure 4). Slide sites visited to establish 'ground truth' included some very large features involving massive rocks carried more than 1 km on, or within, a bed of fine clayey material. They, therefore, have characteristics of 'runout' slides which occur with great rapidity. However, **if** the 1945 event had a recurrence interval of >100 years, then the mosaic of old landslides identifiable on the aerial photographs may represent 10^3 years of landslide activity, and overlap with important changes in climate.

2.3.2. *Sensitivity to climatic change*
From the above example it is inferred that climatic change will be experienced partly through changes to the magnitude and frequency of events, notably rainfall, but also due to drought. In cyclonic areas this will obviously involve wind velocity, but the strength and persistence of Trade Winds of low relative humidity are also implicated over areas such as central and western Africa and much of Southeast Asia. It is, however, difficult to trace the impacts of such changes unless very detailed, dated stratigraphies are available.

Most direct evidence for climatic change on land comes from vegetation changes as recorded in pollen sequences derived from lake sediment or peat swamp cores. This type of evidence tends to indicate periods of drier climate and may also show cooling, as found in the lowland rainforest taxa in west central Africa (Maley 1987; Maley and Livingstone 1983), as well as in the pollen records from mountain areas (Flenley 1979; Hamilton, 1982). Results from such studies are startling, and reveal

TABLE 1. Tentative chronology of Late Quaternary environmental change in the humid tropics

Radiocarbon yr B.P.	Probable environmental conditions and possible phases of forest change
400-700	possible dry period associated with the 'Little Ice age' (1300-1600 A.D.)
1300-1100	climatic crisis in central America (Mayan culture) (700-900 A.D.)
	(pollen signals confused by widespread forest clearance)
3100-2400	possible drier conditions, but strong anthropogenic impact (West Africa; post 2400 in Brazil)
3400-3100	increased humidity in forested tropics with rising discharges, several lesser oscillations of humidity followed
4200-3400	mid Holocene dry phase, probably quite severe (Africa, Brazil)
5500-4200	declining humidity in some areas of the humid tropics (dry excursions in Amazonia, 5500; 4800)
7000-5500	increased humidity and modest rise in lake levels
	significant human occupation and agriculture
7800-7000	reduced lake levels and river discharges in West and East Africa, Brazil
10,500-8000	second humid period, with high lake levels and discharges
	rainforest widely re-established 9000-8000
11,000-10,500	dry, cool interval in many areas; low lake levels (corresponding with the Younger Dryas)
12,500-11,000	rapid warming with unstable climates and prolonged heavy rains in tropical Africa, very high lake levels world-wide
-13,000/12,000	some warming after 15,000; significant changes post 13,000
	many rainforest areas reduced to woodland and savanna mosaics (predicted forest refuges much debated: inner Amazon and Zaire basins; Atlantic margins of Africa; some islands of Southeast Asia)
18,000 (LGM)-	tree-line lowered >1000 m; rainfalls possibly reduced by 30-60% and temperatures depressed by $4^{\circ}C+/-2^{\circ}C$; persistent cooler and drier conditions in many lowland areas for 5-7000 yr
Post 22,000 -	becoming cold in uplands and dry in most lowlands
	earliest human occupation precedes LGM in some areas

Main sources: Absy *et al.* (1989, 1991); Adamson *et al.* (1980); Butzer (1980); Flenley (1988); Gasse and van Campo (1994); Giresse *et al.* (1994); Haberle *et al.* (1991); Hope and Tulip (1994); Jolly *et al.* (1994); Kershaw (1978, 1992); Lezine and Vergnaud-Grazzini (1993); Maloney (1997); Oldfield *et al.* (1985); Schubert (1988); Street and Grove (1979); Street-Perrott and Perrott (1990); Talbot *et al.* (1984); Thomas and Thorp (1980); Thomas *et al.* (1985); Thorp *et al.* (1990); van der Hammen and Absy (1994); van der Hammen *et al.* (1992 a,b).

major shrinkage and fragmentation of tropical rainforest areas during the LGM, persisting possibly for 5000-7000 years. Such major changes could only have come about as a result of climatic change on a global scale and probably involving a depression in annual rainfalls of 30-60% (van der Hammen and Absy 1994; Peters and

Tetzlaff 1990; Heaney 1991; Verstappen 1994), and annual temperatures by 4°+/-2°C (Bonnefille *et al.* 1990; Markgraf 1989; van der Hammen 1991; Vincens *et al.* 1993). Although the changes would have been regionally varied, current knowledge suggests that a large part of the lowland humid tropics experienced magnitudes of change within these limits (temperature reduction 4°+/-2°C, rainfall reduction 30-60%: Table 1).

A recent GCM developed by Dong *et al.* (1996) produced regional averaged precipitation over northern Africa (N of 10° N) for the LGM of -40.62% compared to present day, and +23.21% for the Holocene climatic optimum. Equivalent figures for southern Asia were -64.44 % and 17,93% respectively. Comparable changes in soil moisture were predicted and, perhaps most interesting of all, the wetter climates of the Holocene were shown to have included much higher daily rainfalls and soil moisture levels during the wet season. Such swings of 64-82% in mean rainfall between the driest and wettest periods of the last 21ka, when combined with soil moisture changes, would largely explain the widespread changes in hydrology, erosion and sedimentation observed in tropical landscapes over this period (Table 1).

What such climates were like is less clear, although increases in seasonality, reduction in storm magnitude and frequency due to failure of the monsoons, and low activity in the ITCZ (Inter-tropical Convergence Zone) have all been postulated for the LGM. Some areas in west and central Africa probably experienced an increased intensity of the NE Trades, but lowered temperatures in equatorial regions have been associated with greater cloudiness, and there is evidence for the drying out of lakes (Street-Perrot *et al.* 1985) and a lack of sedimentation in small rivers (Thomas and Thorp 1980, 1992, 1995). It seems likely that forests failed to regenerate over large areas soon after 21,000 B.P., leaving many slopes and hillsides less well protected under a deciduous woodland or savanna. Less prolonged, but still intense, rainfalls may have led to much local sediment transfer on steeper slopes with the accumulation of colluvium. These ideas have been reported elsewhere (Thomas and Thorp 1992, 1995).

Between *c.* 15,000-13,000 B.P., it appears that tropical climates began to change, as part of the global warming at the end of the last ice age. Lake levels began to rise worldwide, but did not reach their maxima until about 10,000 B.P. Around *c.* 12,700-11,000 B.P. many rivers in Africa, including the Nile, experienced high floods and these recurred again after 10,000 B.P.; the intervening, drier, period corresponding with the European Younger Dryas.

Widespread landscape instability appears to have accompanied these changes, but we have little knowledge of the climatic patterns of the period. Many of the sediment stores in contemporary landscapes may date either from the dry late Pleistocene climate or the wetter early Holocene climates. Fans appear to have accumulated in the earlier phases, to become incised as the forest reclaimed the slopes and sediment loads became smaller and finer with increased soil development. When wetter conditions prevailed, landsliding probably increased in severity, possibly reaching a peak during the Holocene Pluvial period between 9000-7000 B.P. After this time, the Holocene climates show similar but short-lived fluctuations in precipitation (and perhaps temperature).

There is some speculation in the scenarios advanced for these periods, yet the central point is clear: that the landscapes of the humid tropics have passed through numerous crises driven by important climatic changes in the past. The amplitude of these changes was considerable, and equivalent to a 6-8° shift in latitude in west Africa (Lezine 1991; Lezine and Casanova 1989). In the outer, more seasonal, parts of the dry lowland rainforest it is likely that the forest disappeared completely, but the possibility of favoured edaphic refuges existed will have increased towards the central parts of the rainforest zone. According to position, different proportions and sites within the forested landscapes would have become sensitive to change (cf. Clapperton 1993; Hamilton and Taylor 1991; Livingstone 1982; Whitmore and Prance 1987).

2.4. SENSITIVITY OF THE QUATERNARY DEPOSITS

The legacy of Quaternary climate change is seen in sediment sequences, often involving buried soils; in palaeosols, and in erosional features which include old landscape scars, and deep fluvial channels now buried by later sediment infill. These palaeochannels are important economically because they often contain rich placer deposits (e.g. diamond, gold, cassiterite, rutile). Some of these channels are now found off-shore and were formed during the lowered global sea level (c. -100 m) of the glacial maxima; others were excavated as a consequence of increased discharges and floods post 12,700 B.P.

This type of inheritance in the landscape is found worldwide, but in the humid tropics these landforms and sediments, with their associated soils are particularly liable to rapid re-adjustment to the prevailing environmental system. In general terms this means that many sites that prove unstable or liable to erosion may represent the temporary storage of sediment which is subsequently shed quite rapidly into river systems. The triggers for this process are often those which have already been considered: namely, the occurrence of extreme rainfall events, including periods of prolonged wetness, and the operation of internal thresholds, as when continued sedimentation in hillslope hollows leads to failure. However, human interference with slope conditions is a potent threat to the stability of all humid tropical landscapes, but particularly where either thick saprolites or inherited deposits are found. In the final section of this discussion some illustrations of these problems will be offered.

2.4.1.   Loss of vegetation cover

Loss of vegetation cover is caused most widely by the extension or intensification of agriculture (including grazing), but may occur locally for wood products without regard to subsequent land use. In addition, people and animals can trigger serious erosion along trackways and at springheads and wells, where water is collected (Grove, 1951, and Figure 5). Knowledge of early settlement and partial clearance of rainforest for agriculture is still very sketchy (see Maloney, this volume). Human settlement in Irian Jaya may have occurred as early as 26,000 B.P. (Haberle et al. 1991), but deforestation in the lowland humid tropics before 5000 B.P. was probably small scale, short-term and spatially very patchy and its impacts may have been short-lived and localised, until widespread agriculture associated with larger populations spread across many tropical areas in a diachronous fashion, perhaps from about 5000 B.P. onwards. However, the main spread was during the last 2000-3000 years (Flenley 1988; Hope and Tulip 1994; Jolly et al. 1994). There is no doubt that the extension of gardens in shifting agricultural systems can lead to increases in sediment yield, but human impact increases markedly with the building of roads and formation of a closed pattern of settlement and fields.

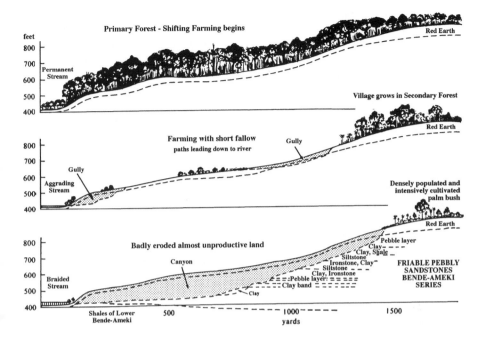

FIGURE 5. Development of canyon-like gullies in southeastern Nigeria as recorded by Grove in 1951. Now probably more than a century old, these features are thought to have enlarged from small gullies around compounds, forest tracks and springheads in a farming landscape. Very deep weathering in Cretaceous sandstones has been a major contributing Grove (1951). Now probably more than a century old, these features are thought to have enlarged from small gullies around compounds, forest tracks and springheads. The very deep weathering in Cretaceous sandstones is a major contributory factor in their subsequent development, and it is possible that these gullies originated in the early Holocene, but we have no direct evidence for this (Grove, pers. comm. 1997). Reproduced with the permission of the author and the Crown Agents.

In their study of three lakes (above 1800 m asl) in Papua New Guinea Oldfield *et al.* (1985) showed that from 500 B.P. until 1973 one undisturbed lake catchment (Pipiak) experienced no significant increase in sedimentation, while another (Ipea) demonstrated a 20-32x increase in sediment entering the lake. Other examples of such effects are found widely in the literature (Spencer and Douglas 1985; Thomas 1994).

The consequences of forest destruction on these substrates is to expose the surface layers to intense rainfall, rainsplash and surface erosion. Prior clay eluviation tends to produce a sandy topsoil, and although this may have a high inherent infiltration capacity, the remaining clay particles are readily washed into the pores, reducing this potential at shallow depths and promoting surface runoff and erosion. Over more basic rocks, on steeper slopes, and in some lower rainfall, marginal forest areas of the humid tropics the surface soil horizons will retain more clay. Once the aggregates are broken down, due to loss of organic matter and exposure to high energy rain drop impact, surface flows are more likely. However, few areas of former rainforest remain bare of vegetation and grasses and shrubs, if not forest regrowth, quickly reclaim most cleared land. Over nutrient rich substrates, such as volcanic lavas and ashes and many basic igneous rocks regrowth may be vigorous.

The sensitive areas in this respect tend to be broad zones of sandy, leached soils over sedimentary formations which dominate many of the basins present or former interior drainage in Africa and South America. Such formations vary in age from the Proterozoic Rokel River Group sediments in Sierra Leone through to the Mesozoic basin sediments of much of NW Africa and the 'Continental Terminal' of Tertiary age, to the Quaternary (including recent) fluvial and aeolian sediments of the great riverine basins of the Congo (and Amazon in South America) and the coastal plain sediments of the Guyana, Kalimantan and many other lowland coasts.

Some of the potential results of exposure of these soils to direct impact of tropical rainfall include:

i.   progressive loss of topsoil from sheet wash, often leading to the formation of gullies and loss of farmland

ii. shallow landsliding and debris flow events during intense rainstorms, wherever slopes exceed 12° (Larsen and Torres-Sanchez 1996) and particularly in steeper areas (20-26° slope)

iii. development of shear failures and slides, often leading to flowage of sediments stored in hillslope hollows, commonly found upslope from stream heads

iv. erosion of fine grained floodplain sediments, as a consequence of exposure to rapid bank erosion, by removal of protective vegetation, construction of unprotected trackways, mine sites and buildings

v.   loss of organic topsoil and possibly increased leaching of coarse sandy sediments in *white sand* podzolic soil environments, in riverine or coastal zone locations, usually as a result of deforestation and leading to localised, cut-over scrubland.

The downstream effects of this sensitivity to erosion will be a loss of quality in river water leading to problems for domestic and animal drinking supplies, irrigation, fish stocks, and dam siltation. This emphasises another aspect of sensitivity in these environments because most rivers in non-mountainous, forested catchments carry a low suspended sediment load. But the widespread availability of fine grained floodplain sediments and slope materials able to enter the stream channel, under the conditions outlined above, will lead to very rapid increases in sediment load. Many figures exist to describe this impact, but they can be summarised by reference to Bruijnzeel (1983) who observed that 0.2 t Ha^{-1} y^{-1} characterised the sediment yield from slopes under rainforest, while under cultivation or as a result of clearance this can rise to 600-1200 t Ha^{-1} y^{-1} (cf. Lal *et al.* 1981, Lal 1986).

2.4.2.   Loss of basal support
The removal of basal support during excavation for roads and other construction is a potent source of landscape instability in all environments, but has a particular context in the humid tropics:

   i.   failure and of saprolite and colluvium stored on steep slopes is
         triggered by prolonged rainfalls, as in Hong Kong (Lumb 1975)

   ii. reactivation of palaeolandslides by renewed failure along old shear planes
         where these become exposed in cuttings can result in major landslides

   iii. development of shear failures along discontinuities in unconsolidated
         materials such as slope sediments and weathered volcanic deposits which
         give rise to frequent slope failures along roads, as in Java (Brand *et al.*
         1985).

This also serves to highlight a further application of sensitivity studies, to volcanic landscapes, where stratified ash deposits form thick sequences which weather very rapidly in humid tropical regions. Where dominated by allophane or halloysite, these materials are free draining and can remain stable under cultivation on very steep slopes >45°. However, they may contain vermiculite and smectite clays which absorb large quantities of water and can be stratified and contain impermeable layers, so that confinement of the water, combined with swelling properties of the clay lead to frequent mudflows, a frequent phenomenon throughout the island of Java, for example (Heath and Sarosso 1988).

2.5. MODELS DESCRIBING LANDSCAPE SENSITIVITY

Models designed to describe landscape sensitivity to climatic change have mainly been developed to address loss of vegetation in highly seasonal climates (Langbein and Schumm 1958; Kirkby 1978; Knox 1972, 1984; Roberts and Baker 1993). They embody two observations. Firstly, sediment yield is thought to rise with annual rainfall in semi-arid regions until sufficient moisture is available to nourish a protective plant cover; runoff and sediment yield then both decline steadily as humid climatic conditions are approached, both values reaching a minimum in the forest climates. Secondly, disturbance of the vegetation cover produces an abrupt response in terms of increased runoff and erosion on slopes, and a 'flashy' regime in rivers, which experience increased flood peaks and reduced low flows.

These models are difficult to apply to rainforest regions, however, because where rainfalls may average around 2000 mm yr^{-1}, reductions of 30-60% will not cross a threshold of maximum runoff (around 700 mm yr^{-1}, according to Schumm (1965)). Additionally, Roose (1981) suggests that rainfall *erosivity* declines with annual rainfall in a linear fashion. As climate becomes gradually drier, therefore, we might anticipate that no abrupt changes in sediment yields will occur. The tree cover may not regenerate in its original form, but other species adapted to cooler and/or drier conditions will invade the forest without interruption. Gradually, a more open plant cover may appear on sites where drought stress is most severe, and local transfers of sediments may increase on exposed hillsides, leading to accumulation of sediment stores along valley sides as colluvium.

Evidence suggests that the prolonged aridity of the late Pleistocene rendered many rivers inactive and a few floodplain sediments date from this period. The rapid warming of climate at the termination of the last ice age, however, would have been very different. Increased storminess and rainfall are likely to have had a major impact on slopes adjusted to the drier conditions, leading to increased rates of fan formation, and subsequently to the trenching of the fans and to major sedimentation in the river floodplains.

It appears that the forest took 2-3 millennia to recover completely, after which hillslope processes may have become increasingly dominated by landslides and soil flows during the Holocene pluvial. From around 8000 B.P. onwards it becomes less easy to decipher the record because of increased human impact on the vegetation (Table 1). But, although the oscillations of climate did reduce in amplitude, Knox (1993) has found in the USA that quite small changes to external climate can be amplified in the record of stream activity, and phases of sedimentation continued throughout the mid-Holocene (Thomas and Thorp 1995).

It is not possible, therefore, to adopt existing process-response models without modification. However, because of the frequency with which complex alluvial stratigraphies are encountered (which mostly appear to be Holocene in age), it appears that repeated landscape instability has resulted from the small shifts in many rainforest climates during the last 8000 years.

3.   Possible effects of future climatic change

Predictions of possible future climatic change for tropical rainforest areas as a result of global warming remain very uncertain, and the most recent IPCC appraisal concludes that, 'tropical forests are likely to be more affected by changes in land use than by climatic change as long as deforestation continues at its current high rate' (Kirschbaum and Fischlin 1996, p. 97). Loss of rainforest from large areas such as the Amazon Basin as a result of clearance may, however, reduce rainfall, and Salati and Vose (1984) have suggested that widespread deforestation (c. 50%) could lead to a 20% reduction in precipitation. The IPCC (Kirschbaum and Fischlin 1996) admit that the transient response of tropical rainforests to a 1°C increase in temperature due to global warming is unknown at the regional scale, and that most effects will come from *water availability.* Recently, Rhind (1995) has argued against the view that global warming will lead to increased cloudiness in the humid tropics, suggesting that reductions in low cloud would be accompanied by increased precipitation and that a more vigorous atmospheric circulation would fuel more cyclonic storms in hurricane-prone areas. Forest loss may arise in marginal areas, where reductions in rainfall seem possible.

Increases in rainfall may lead to more throughfall and to enhanced runoff, but soil moisture is also likely to rise and a higher frequency of landsliding may ensue. Thus rates of erosional processes and sediment yield will rise. Forest degradation and an increased incidence of fires during runs of dry years can also have disasterous effects, as deep saprolites and fragile colluvia are evacuated from the landscape by overland flow. Such changes have been conceptualised as a simplified model of how changing climatic conditions may disturb rainforest equilibria in Figure 6. Climate modellers, using sea surface temperature data from ocean cores, have in the past ignored empirical evidence for major environmental changes on land in the humid tropics during the last glacial maximum (Rhind 1995; Thomas 1994; Thomas and Thorp 1995). It would be a grave mistake to use the paucity of evidence on some of these issues to infer that global warming will have little impact on the tropical rainforests in the future. The possible impact of rainforest destruction on local, regional and global climates is discussed further in McGuffie *et al.*, this volume.

4. Conclusion

Landscape sensitivity in the tropical forest lands, as elsewhere, has been shown to have both strong spatial heterogeneity and also to be subject to temporal fluctuation on different timescales. In the face of continuing pressure on the forested areas of the tropics from agricultural extension and timber exploitation, conservation policies need to embrace more than existing systems for soil conservation, or improved silvicultural methods. Each relatively homogenous natural area, or landscape element, possesses a unique combination of climate, rocks, landforms and soils, and the need for good local knowledge of these features should not be overridden by uncritical general statements. Nevertheless, the development and application of general principles to the problems presented by land clearance in the humid tropics must clearly also be attempted.

One context of such a study now has to be the potential impacts of future climatic change (see McGuffie *et al.*, this volume). In this respect we often have better knowledge of how landscapes reacted to past climatic changes than we do of the possible future behaviour of contemporary environmental systems, based on the modelling of system parameters alone. Ideally such approaches need to be brought together, but this has not so far been achieved. Furthermore, the overlap between human settlement and climate shifts on shorter (10^2-10^3 y) timescales leads to a confusion of signals. The case of the collapse of the Mayan civilisation has been a subject of much discussion; recently the importance of climatic change between 1300-1100 B.P. (A.D. 700-900) as a factor leading to its demise has been stressed (Hodell *et al.* 1995).

Present understanding of some of these problems suggests that a magnitude and frequency approach to climate and climatic change is necessary to understand sensitivity to erosion and sedimentation in the landscape (Ahnert 1987; Eybergen and Imeson 1989). This involves timescales from the occurrence of glacial cycles (10^5 y) to the incidence of rainfall on a daily, possibly even hourly, basis. Our present state of

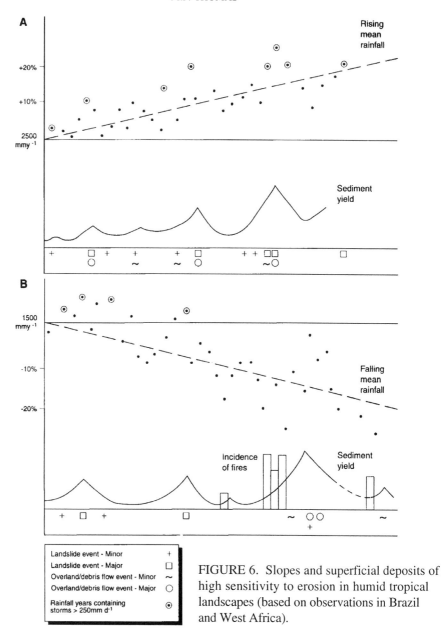

FIGURE 6. Slopes and superficial deposits of high sensitivity to erosion in humid tropical landscapes (based on observations in Brazil and West Africa).

A. Slopes in deep saprolite may erode to produce bottle-shaped gullies [A], or fail to form landslides [B] and colluvium [C] which may extend into deeply alluviated valleys [D] which may also become deeply gullied, typified by the Bananal region of E. Brazil (Coelho Netto *et al.* (1990), de Oliviera (1991)).

B. Slopes underlain by thin regoliths often exhibit shallow debris slides [A] or infilled palaeogullies [B].

understanding of past climates does not allow this linkage to be completed. While climate variability and change are presently understood on scales from decades to millennia and beyond (Williams *et al.* 1994), it is only with respect to the former that the incidence of daily rainfalls can be analysed.

The significance of sensitivity studies for future landscape stability and the sustainability of existing or novel land use practices requires more emphasis and much greater analytical detail. For rainforest areas climate change scenarios are not always clear and much more attention has been paid to the drier parts of the seasonal tropics. But, if parts of the inner tropics become wetter, this may stimulate conditions during the Holocene 'pluvial' between 10,000-8000 B.P., when increased flooding and landslide activity typified many non-cyclonic areas. In both these cases, although we cannot be certain of future outcomes, the need for rainforest conservation in sensitive areas becomes urgent and of paramount importance. Equally, the careful management of cleared areas, whether for tree planting (see Barrow and Fraser, this volume), field crops or for construction demands that attention be paid to the geological and geomorphological components of landscape sensitivity. Particularly where surficial deposits are deep, it is important to think beyond the conventions of soil conservation to consider the protection of the total land resource which can be destroyed by combinations of landsliding, gullying, flooding and sedimentation.

References

Absy, M.L., Cleef, A, Fournier, M., Martin, L, Servant, M., Siffedine, A, Ferreira da Silva, M., Soubies., F, Suguio, K., Turcq, B. and Van der Hammen, T. (1991). Mise en evidence de quatre phases d'ouverture de la foret dense dans le sud-est de l'Amazonie au cours des 60 000 dernieres annees. Premiere comparaison avec d'autres regions tropicales. *Comptes Rendus de l'Academie des Sciences*, Paris, 312, Serie II, 673-678.

Absy, M.L., Van der Hammen, T., Soubies., F, Suguio, K., Martin, L., Fournier, M.and Turcq, B. (1989). Data on the history of vegetation and climate in Carajas, eastern Amazonia, *International Symposium on Global Changes in South America During the Quaternary, Sao Paulo, 1989, Special Publication*, 1, 129-131, Sao Paulo, Brazil.

Ab'Saber, A. N. (1988). A Serra do Mar na Regiao de Cubatao: avalanches de Janeiro de 1985, in A.N. Ab'Saber (ed.), *A Rup. Ecologico Na Serra De Paranapiacaba E A Policao Industrial*, Brasil, pp. 74-116.

Adamson, D.A., Gasse, F., Street, F.A. and Williams, M.A.J. (1980). Late Quaternary history of the Nile. *Nature*, 287, 50-55.

Ahnert, F. (1987). An approach to the identification of morphoclimates, in V.G. Gardiner (ed.), *International Geomorphology 1986*, pp. 159-188.

Bigarella, J.J. and Becker, R.D. (eds.) (1975). International symposium on the Quaternary (Southern Brazil, July,15-31, 1975), *Boletim aranaense de Geociencias*, 33, 1- 370.

Bonnefille, R., Roeland, J.C. and Guiot, J. (1990). Temperature and rainfall estimates for the past 40,000 years in equatorial Africa, *Nature*, 346, 347-349.

Brand, E.W., Premchitt, J. and Phillipson, H.B. (1985). Relationship between rainfall and landslides in Hong Kong, *IV International Symposium on Landslides*, Toronto 1984, Vol 1, 377-384, Toronto.

Bruijnzeel, L.A. (1983). Evaluation of runoff sources in a forested basin in a wet monsoonal environment: a combined hydrological and hydrochemical approach, in E. Keller (ed.), *Hydrology of Humid Tropical Regions*, IAHS Publication, 140, 165-174.

Brunig, E.F. (1974). *Ecological Studies in the Kerangas Forests of Sarawak and Brunei*. Borneo Literature Bureau, Kuching, Sarawak, Malaysia.

Brunig, E.F. (1975). Tropical ecosystems: state and targets of research into the ecology of humid tropical ecosystems, *Plant Research and Development*, **1**, 22-38.

Brunsden, D. and Thornes, J.B. (1979). Landscape sensitivity and change, *Transactions of the Institute of British Geographers*, n.s.**4**, 463-484.

Brunsden, D. (1980). Applicable models of long term landform evolution, in H. Hagedorn and M.F. Thomas (eds.), Perspectives in Geomorphology, *Zeitshrift fur Geomorphologie*, Supplbd. **36**, 16-26.

Butzer, K.W. (1980). Pleistocene history of the Nile Valley in Egypt and Lower Nubia, in M.A.J. Williams and H. Faure (eds.), *The Sahara and the Nile*, Balkema, Rotterdam, pp. 253-280.

Caine, N. (1980). The rainfall intensity-duration control of shallow landslides and debris flows, *Geografiska Annaler*, **62 A**, 23-27.

Chauvel, A, Lucas, Y, and Boulet, R. (1987). On the genesis of the soil mantle of the region of Manaus, Central Amazonia, Brazil, *Experientia*, **43**, 234-241.

Clapperton, C.M. (1993). Nature of environmental changes in South America at the Last Glacial Maximum, *Palaeogeography, Palaeoclimatology ,Palaeoecology*, **101**, 189-208.

Coellho Netto, A.L. and Fernandes, N.F. (1990). Hillslope erosion, sedimentation and relief inversion in SE Brazil: Bananal, SP., R.R. Zeimer, C.L. O'Loughlan and L.S. Hamilton (eds.), *Research Needs and Applications to Reduce Erosion and Sedimentation in Tropical Steeplands*, IAHS Publication No. **192**, 174-182.

Deere, D.U. and Patton, F.D. (1971). Slope stability in residual soils. *Proceedings of the 4th Panamerican Conference on Soil Mechanics and Foundation Engineering*, Caracas, June 1971, 1, pp. 87-170, American Society of Civil Engineers, New York.

De Moura, J.R. da S., Da Silva, T.M., Mello, C.L., Peixoto, M.N. de O., Santos, A.A. de M. and Esteves, A.A. (1989). Roteiro geomorphologico-estratigrafico da regiao de Bananal (Sao Paulo), *Publicacao Especial No. 2. pp. 28. 2o Congresso da Associacao Brasileira de Estudos do Quaternario, 1989*, Universidade Federal do Rio De Janeiro, Brazil.

Dietrich, W.E. and Dorn, R. (1984). Significance of thick deposits of colluvium on hillslopes: a case study involving the use of pollen analysis in the coastal mountains of California, *Journal of Geology*, **92**, 147-158.

Dietrich, W.E., Wilson, C.J. and Reneau, S.L. (1986). Hollows, colluvium and landslides in soil-mantled slopes, in A.D. Abrahams (ed.), *Hillslope Processes* (The Binghamton Symposia in Geomorphology: International Series, no 16), Allen and Unwin, Boston and London, pp. 361-388.

Dong, B., Valdes, P.J. and Hall, N.M.J. (1996). The changes of monsoonal climates due to Earth's orbital perturbations and Ice Age boundary conditions, *Palaeoclimates*, **1**, 203-240.

Douglas, I. (1967b). Man, vegetation and the sediment yield of rivers, *Nature*, **215**, 925-928.

Douglas, I. (1973). *Rates of Denudation in Selected Small Catchments in Eastern Australia*, University of Hull Occasional Papers in Geography, 21, Hull.

Downs, P.W. and Gregory, K.J. (1993). The sensitivity of river channels in the landscape system, in D.S.G. Thomas and R.J. Allison (eds.), *Landscape Sensitivity*, John Wiley and Sons, Chichester, pp. 15-30.

Duchaufour, P. (1982). *Pedology, Pedogenesis and Classification* (English ed., trans. T.R. Paton), George Allen and Unwin, London

Ellen, S.D.,and Wieczorek, G.F. (eds.) *Landslides, Floods, and Marine Effects of the Storm of January 3-5, 1982, in the San Francisco Bay Region, California*. U.S. Geological Survey Professional Paper, **1434**, pp. 310, Washington.

Eybergen, F.A. and Imeson, A.C. (1989). Geomorphological processes and climatic change, *Catena, 16*, 307-319.

FAO (1968). *Soil and Land Use Survey of Part of the Eastern Province, Sierra Leone* (based on the work of J. Stark), Report No. TA 2574 to The Government of Sierra Leone, FAO, Rome.

Ferreira, R.C. and Monteiro, L.B. (1985). Identification and evaluation of collapsibility of colluvial soils that occur in the Sao Paulo State. *First International Conference on Geomechanics in Tropical Lateritic and Saprolitic Soils*, Brasilia, 1985, Volume 1, pp. 269-280.

Flenley, J.R. (1979). *The Equatorial Rainforest: a Geological History*, Butterworth, London.

Flenley, J.R. (1988). Palynological evidence for land use changes in South-East Asia, *Journal of `Biogeography*, **15**, 185-197.

Garwood, N.C., Janos, D.P. and Brokaw, N. (1977). Earthquake-caused landslides: a major disturbance to tropical forests, *Science*, **205**, 997-999.

Gasse, F. and Van Campo, E. (1994). Abrupt post-glacial climatic events in West Asia and North Africa monsoon domains, *Earth and Planetary Science Letters*, **126**, 435-456.

Geological Society (1990). Engineering Group Working Party Report: Tropical residual soils, *Quarterly Journal of Engineering Geology*, **23**, 1-101.

Giresse, P.; Maley, J. and Brenac, P. (1994). Late Quaternary palaeoenvironments in the Lake Barombi Mbo (West Cameroon) deduced from pollen and carbon isotopes of organic matter, *Palaeogeography, Palaeoclimatology, Palaeoecology*, **107**, 65-78.

Grove, A.T. (1951). *Land Use and Soil Conservation in Parts of Onitsha and Owerri Provinces*, Geological Survey of Nigeria, Bulletin, 21, Kaduna, Nigeria.

Haberle, S.G., Hope, G.S. and Defretes, Y. (1991). Environmental change in the Baliem Valley, montane Irian Jaya, Republic of Indonesia, *Journal of Biogeography*, **18**, 25-40.

Hamilton, A.C. (1982). *Environmental History of East Africa: A Study of the Quaternary*, Academic Press, New York.

Hamilton, A.C. and Taylor, D. (1991). History of climate and forests in tropical Africa during the last 8 million years, *Climate Change*, (Special Issue: Tropical Forests and Climate), **19**, 65-78.

Heaney, L.R. (1991). A synopsis of climatic and vegetation change in southeast Asia, *Climate Change*, (Special Issue: Tropical Forests and Climate), **19**, 53-61.

Heath, W. and Saroso, B.S. (1988). Natural slope problems related to roads in Java, Indonesia, *Geomechanics in Tropical Soils*, Proceedings of the Second International Conference on Geomechanics in Tropical Soils, Singapore, 1988, Vol. 1, pp. 259-266, Balkema, Rotterdam.

Heyligers, P.C. (1963). *Vegetation and soil of a white-sand savanna in Suriname*, Verhandelingen der koninklinke Nederlandse, Akademie van Wetenschappen, Afd. Natuurkunde. N.V. Noord-Hollandschc Uitgevers Maatschappij, Amsterdam.

Hodell, D.A., Curtis, J.H. and Brenner, M. (1995). Possible role of climate in the collapse of classic Maya civilisation, *Nature*, **375**, 391-394.

Hope, G. and Tulip, J. (1994). Along vegetation history from lowland Irian Jaya, Indonesia, *Palaeogeography, Palaeoclimatology, Palaeoecology*, **109**, 385-398.

Iida, T. and Okunishi, K. (1983). Development of hillslopes due to landslides, *Zeitschrift für Geomorphologie*, n.f. Supplbd, **46**, 67-77.

Jibson, R.W. (1989). Debris flows in southern Puerto Rico, in J.P. Schultz and R.W. Jibson (eds.), *Landslide Processes of the Eastern United States and Puerto Rico*, Geological Society of America, Special paper 236, 29-55.

Jolly, D., Bonnefille, R. and Roux, M. (1994). Numerical interpretation of a high resolution Holocene pollen record from Burundi,*Palaeogeography, Palaeoclimatology, Palaeoecology*, 109, 357-370.

Jones, F.O. (1973). Landslides of Rio de Janeiro and the Serra das Araras Escarpment, Brazil. *US Geological Survey Professional Paper*, **697**, pp. 42.

Kershaw, A.P. (1978). Record of last interglacial-glacial cycle from northeastern Queensland, *Nature*, **272**, 159-161.

Kershaw, A.P. (1992). The development of rainforest-savanna boundaries in tropical Australia, in P.A. Furley, J. Proctor and J.A. Ratter (eds.), *Nature and Dynamics of Forest-Savanna Boundaries*, Chapman and Hall, London, pp. 255-271.

King, J., Loveday, I., and Schuster, R.L. (1989). The 1985 landslide dam and resulting debris flow, Papua New Guinea, *Quarterly Journal of Engineering Geology*, **22**, 257-270.

Kirkby, M.J. (ed.) (1978). *Hillslope Hydrology*, Wiley, Chichester.

Kirschbaum, M.U.F. and Fischlin, A. (eds.) (1996). Climate change impact on forests, in R.T. Watson, M.C. Zinyowera and R.H. Moss (eds). *Climate Change 1995*. Contribution of Working Group II to second Assessment Report of IPCC. Cambridge University Press, Cambridge, pp. 95-129.

Knox, J.C. (1972). Valley alluviation in southwestern Wisconsin, *Annals of the Association of American Geographers*, **62**, 401-410.

Knox, J.C. (1984). Responses of river systems to Holocene climates, in S.C. Porter and H.E. Wright (eds.), *Late Quaternary Climates of the United States* Vol. 2, University of Minnesota Press, Minneapolis, pp. 26-41.

Knox, J. C. (1993). Large increases in flood magnitude in response to modest changes in climate, *Nature*, **361**, 430-432.

Lal., R. (1986). Deforestation and soil erosion, in R. Lal, P.A.Sanchez and R.W. Cummings, Jr. (eds.), *Land Clearing and Development in the Tropics*, Balkema, Rotterdam, pp. 299-315.

Lal, R., Lawson, T.L. and Anastase, A.H. (1981). Erosivity of tropical rains, in M. De Broodt and D.Gabriels (eds.), *Assessment of Erosion*, Wiley, Chichester, pp. 143-151.

Langbein, W.B. and Schumm, S.A. (1958). Yield of sediment in relation to mean annual precipitation, *Transactions of the American Geophysical Union*, **39**, 1076-1084.

Larsen, M.C. and Simon, A. (1993). A rainfall intensity-duration threshold for landslides in a humid tropical environment, Puerto Rico, *Geografiska Annaler*, **75A**, 13-23.

Larsen, M.C. and Torres-Sanchez, A.J. (1996). *Geographic relations of landslide distribution and assessment of landslide hazards in the Blanco, Cibuco, and Coama basins, Puerto Rico*, U.S. Geological Survey Water-Resources Investigations report 95-4029, 56pp, San Juan, Puerto Rico.

Leigh, C.H. (1982). Sediment transport by surface wash and throughflow at the Pasoh Forest Reseerve, Negri Sembilan, Peninsular Malaysia, *Geografiska Annaler*, **64A**, 171-180.

Lezine, A-M. (1991). West African paleoclimates during the last climatic cycle inferred from an Atlantic deep-sea pollen record, *Quaternary Research*, **35**, 456-463.

Lezine, A-M. and Casanova, J. (1989). Pollen and hydrological evidence for the interpretation of past climates in tropical west Africa during the Holocene, *Quaternary Science Reviews*, **8**, 45-55.

Lezine, A-M. and Vergnaud-Grazzini, C. (1993). Evidence of forest extension in West Africa since 22,000 BP: a pollen record from the eastern tropical Atlantic, *Quaternary Science Reviews*, **12**, 203-210.

Livingstone, D.A. (1982). Quaternary geography of Africa and the Refuge Theory, in G.T. Prance (ed.), *Biological Diversification in the Tropics*, Columbia, New York, pp. 523-536.

Loffler, E. (1974). Piping and pseudokarst features in the tropical lowlands of New Guinea, *Erdkunde*, **28**, 13-18.

Loffler, E. (1977). *Geomorphology of Papua New Guinea*, CSIRO/Australian National University Press, Canberra.

Lucas, Y., Boulet, R., Chauvel, A. and Veillon, L. (1987). Systemes sols ferrallitiques-podzols en region amazonienne, in D. Righi and A. Chauvel (eds), *Podzols et Podzolisation*, Comptes Rendus de la Table Ronde International, 1986, Association Francaise pour l'etude du Sol, INRA/ORSTOM, Poitiers, pp. 53-65.

Lucas, Y., Boulet, R., and Chauvel, A. (1988). Intervention simultanee des phenomenes d'enforcement vertical et de transformation laterale dans la mise en place de systemes sols ferrallitiques-podzols de l'Amazonie Brasilienne, *Comptes Rendus l'Acadmie des Sciences*, Paris, Serie 2, **306**, 1395-1400.

Lumb, P. (1975). Slope failures in Hong Kong, *Quarterly Journal of Engineering Geology*, **8**, 31-65.

Maley, J. (1987). Fragmentation de la foret dense humide Africaine et extension des biotopes montagnards au Quaternaire recent: nouvelles donnees polliniques et chronologiques. Implications paleoclimatiques et biogeographiques, *Palaeoecology of Africa*, **18**, 307-334.

Maley, J. and Livingstone, D.A. (1983). Extension d'un element montagnard dans le sud du Ghana (Afrique de l'Ouest) au Pleistocene superieur et a l'Holocene inferieur: premieres donnees polliniques, *Comptes Rendus de l' Academie des Sciences*, Paris, Serie 2, **296**, 1287-1292.

Maloney B.K. (1997). The long-term history of human activity and rainforest development (this volume).

Markgraf, V. (1989). Palaeoclimates in central and south America since 18000 BP based on pollen and lake-level records, *Quaternary Science Reviews*, **18**, 1-24.

Nanson, G.C. (1986). Episodes of vertical accretion and catastrophic stripping: a model of disequilibrium floodplain development., *Bulletin of the Geological Society of America*, **97**, 1467-1475.

Nilsen, T.H. and Turner, B.L. (1975). Infuence of rainfall and ancient landslide deposition recent landslides (1950-71) in urban areas of Contra Costa County, California, *Geological Survey Bulletin*, **1388**, pp 18.

Nye, P.H. (1954). Some soil-forming processes in the humid tropics. Part I: A field study of a catena in the West African forest, *Journal of Soil Science*, **5**, 7-27.

Oldfield, F., Worsley, A.T. and Appleby, P.G. (1985). Evidence from lake sediments for recent erosion rates in the highlands of Papua New Guinea, in I. Douglas and T. Spencer (eds), *Environmental Change and Tropical Geomorphology*, Allen and Unwin, London, pp. 185-196,.

Ollier, C.D. and Pain C. (1996). *Regolith, Soils and Landforms*. John Wiley and Sons, Chichester.

Peters, M. and Tetzlaff, G. (1990). West African palaeosynoptic patterns at the Last Glacial Maximum, *Theoretical and Applied Climatology*, **42**, 67-79.

Reading, A., Thompson, R.D. and Millington, A.C. (1995). *Humid Tropical Environments*, Longman, London.

Reneau, S.L. and Dietrich, W.E. (1987). Size and location of colluvial landslides in a steep forested landscape, in R.L. Beschta, T. Blinn, C.E. Grant, F.J. Swanson and G.G. Ice (eds.), *Erosion and Sedimentation in the Pacific Rim*, IAHS Publication No 165, 39-48.

Reneau, S.L., Dietrich, W.E., Dorn, R.I., Berger, C.R. and Rubin, M. (1986). Geomorphic and paleoclimatic implications of latest Pleistocene radiocarbon dates from colluvium-mantle hollows, California, *Geology*, **14**, 655-658.

Rhind, D. (1995). Drying out the tropics, *New Scientist*, 6 May 1995, 36-40.

Righi, D and Chauvel, A. (eds) (1987). *Podzols et Podzolisation*, Comptes Rendus de la Table Ronde International, 1986, Association Francaise pour l'etude du Sol, INRA/ORSTOM, Poitiers.

Roberts, N. and Baker, P. (1993). Landscape stability and bio-geomorphic response to past and future climatic shifts in intertropical Africa, in D.S.G. Thomas and R.J. Allison (eds.), *Landscape Sensitivity*, John Wiley and Sons, Chichester, pp. 65-82.

Roose, E.J. (1981). Approach to the definition of rainfall erosivity and soil erodibility in West Africa, in M. de Broodt and D. Gabriels (eds.), *Assessment of Erosion*, Wiley, Chichester.

Salati, E. and Vose, P.B. (1984). Amazon Basin: a system in equilibrium. *Science*, **225**, 129-138.

Schubert, C. (1988). Climatic changes during the last glacial maximum in northern South America and the Caribbean: a review, *Interciencia*, **13**, 128-137.

Schumm, S.A. (1977). *The Fluvial System*, Wiley, New York.

Schumm, S.A. and Parker, R.S. (1973). Implications of complex response of drainage systems for Quaternary alluvial stratigraphy, *Nature*, **243**, 99-100.

Simonett, D.S. (1967). Landslide distribution and earthquakes in the Bewanni and Toricelli Mountains, New Guinea, a statistical analysis, in J.N. Jennings and J.A. Mabbutt (eds.), *Landform Studies from Australia and New Guinea*, Cambridge University Press, Cambridge, pp. 64-84.

Singh, J, Wapakala, W.W. and Chebosi, P.K. (1984). Estimating groundwater recharge based on infiltration characteristics of layered soil. *Challenges in African Hydrology and Water Resources* (Proceedings of the Harare Symposium, 1984), IASH Publication, **144**, 37-45.

Smith, B.J. and de S. Sanchez, B. A. (1992). Erosion hazards in a Brazilian suburb, *Geography Review*, **6**, 37-41.

Spencer, T and Douglas, I. (1985). The significance of environmental change: diversity, disturbance and tropical ecosystems, in I. Douglas and T. Spencer (eds.), *Environmental Change and Tropical Geomorphology*, Allen and Unwin, London, pp. 13-33.

Spencer, T., Douglas, I., Greer, T. and Sinun, W. (1990). Vegetation and fluvial geomorphic processes in south-east Asian tropical rainforests, in J.B. Thornes (ed.), *Vegetation and Erosion*, Wiley, Chichester, pp. 420-436.

Street, F.A. and Grove, A.T. (1979). Global maps of lake level fluctuations since 30,000 BP, *Quaternary Research*, **12**, 83-118.

Street-Perrott, F.A. and Perrott, R.A. (1990). Abrupt climatic fluctuations in the tropics: the influence of Atlantic Ocean circulation, *Nature*, **343**, 607-612.

Street-Perrott, F.A., Roberts, N. and Metcalfe, S. (1985). Geomorphic implications of late Quaternary hydrological and climatic changes in the Northern Hemisphere tropics, in I. Douglas and T. Spencer (eds.) *Environmental Change and Tropical Geomorphology*, Allen and Unwin, London, pp. 164-183.

Talbot, M.R., Livingstone, D.A., Palmer, P.G., Maley, J., Melack, J.M., Delibrias, G. and Gulliksen, S. (1984). Preliminary results from sediment cores from lake Bosumtwi, Ghana, *Palaeoecology of Africa*, **16**, 173-192.

Temple, P.H. and Rapp, A. (1972). Landslides in the Mgeta area, western Uluguru Mountains, Tanzania, *Geografiska Annaler*, **54 A**, 157-193.

Thomas, D.S.G. and Allison, R.J. (1993). *Landscape Sensitivity*. John Wiley and Sons, Chichester.

Thomas, M.F. (1983). Contemporary denudation systems and the effects of climatic change in the humid tropics - some problems from Sierra Leone, in D.J. Briggs and R.S. Waters (eds.), *Studies in Quaternary Geomorphology*, Geo Books, Norwich, pp. 195-214.

Thomas, M.F. (1994). *Geomorphology in the Tropics*. John Wiley and Sons, Chichester.

Thomas, M.F. and Thorp, M.B. (1980). Some aspects of the geomorphological interpretation of Quaternary alluvial sediments in Sierra Leone, *Zeitschrift fur Geomorphologie*, n.f., Supplbd **36**, 140-161.

Thomas, M.F. and Goudie, A.S. (eds.) (1985). Dambos: small channelless valleys in the tropics, *Zeitschrift fur Geomorphologie*, Supplbd, **52**, pp. 222.

Thomas, M.F. and Thorp, M.B. (1992). Landscape dynamics and surface deposits arising from late Quaternary fluctuations in the forest-savanna boundary, in P.A. Furley, J. Proctor and J.A. Ratter (eds.), *Nature and Dynamics of Forest-Savanna Boundaries*, Chapman and Hall, London , pp. 215-253.

Thomas, M.F., Thorp, M.B. and Teeuw, R.M. (1985). Palaeogeomorphology and the occurrence of diamondiferous placer deposits in Koidu, Sierra Leone, *Journal of the Geological Society*, **142**, 789-802.

Thorp, M.B., Thomas, M.F., Martin, T. and Whalley, W.B. (1990). Late Pleistocene sedimentation and landform development in western Kalimanatan (Indonesian Borneo), *Geologie en Mijnbouw*, **69**, 133-150.

Van der Hammen, T. (1972). Changes in vegetation and climate in the Amazon basin and surrounding areas during the Pleistocene, *Geology en Mijnbouw*, **51**, 641-643.

Van der Hammen, T. (1991). Palaeoecological background: neotropics, *Climate Change*, **19**, 37-47 (Special Issue: Tropical Forests and Climate).

Van der Hammen, T. and Absy, M.L. (1994). Amazonia during the last glacial, *Palaeogeography, Palaeoclimatology, Palaeoecology*, **109**, 247-261.

Van der Hammen, T., Duivenvoorden, J.F., Lips, J.M., Urrego, L.E.and Espejo, N. (1992a). Late Quaternary of the middle Caquet· River are (Colombian Amazonia), *Journal of Quaternary Science*, **7**, 45-55.

Van der Hammen, T., Urrego, L.E., Espejo, N., Duivenvoorden, J.F., and Lips, J.M. (1992b). Late-glacial and Holocene sedimentation and fluctuations of river water level in the Caquet· River area (Colombian Amazonia), *Journal of Quaternary Science*, **7**, 57-67.

Vincens, A., Chalie, F., Bonnefille, R., Guiot, J. and Tiercelin J.-J. (1993). Pollen-derived rainfall and temperature estimates from Lake Tanganyika and their implication for late Pleistocene water l evels, *Quaternary Research*, **40**, 343-350.

Watson, R.T.; Zinyowera, M.C. and Moss, R.H. (eds.) (1996). *Climate Change 1995*. Contribution of Working Group II to second Assessment Report of IPCC. Cambridge University Press, Cambridge.

Whitmore, T.C. and Prance, G.T. (eds.) (1987). *Biogeography and Quaternary History in Tropical America*, Clarendon, Oxford.

Williams, M.A.J. (1994). Some implications of past climatic changes in Australia, *Transactions of the Royal Society of Australia*, **118**, 17-25.

Verstappen, H. T. (1994). Climatic change and geomorphology in south and south-east Asia, *Geo-Eco-Trop* **16**, 101-147.

Professor Michael F. Thomas, Department of Environmental Science, University of Stirling, Stirling FK9 4LA, Scotland

3. HUMAN ACTIVITY AND THE TROPICAL RAINFOREST: ARE THE SOILS THE FORGOTTEN COMPONENT OF THE ECOSYSTEM ?

Stephen Nortcliff

1. Introduction

The humid tropical rainforest ecosystem is a pristine environment and serves as an enormous reservoir of sequestered carbon, biological diversity and a wide array of plants and animals, in addition to being a resource of food, timber, medicine and other products for people. One of the major global concerns is the continuing pace of deforestation of this ecosystem which results in the loss of many of the resources (FAO, 1983). In addition many of the land use practices introduced in the post clearance phase have been seen to fail, often within very short time spans following the initial clearance. Deforestation may occur for a variety of reasons including:

    i. Exploitation of the timber;
    ii. Exploitation of other forest products;
    iii. Clearance to provide access to land for agricultural production (arable crops and grazing);
    iv. Clearance for timber plantation;
    v. Clearance for other plantations (e.g. cocoa, rubber, etc.);
    vi. Clearance to provide land for urban and industrial development;
    vii. Clearance to provide access to below ground mineral resources.

These deforestation activities vary to a considerable degree in the extent to which they disrupt the soil system, with major irreversible damage occurring where clearance is to exploit minerals below the soil surface or to facilitate urban or industrial development. Timber exploitation, if poorly managed, may have very damaging effects on the soil system, but where timber exploitation involves selective harvesting coupled with sustainable management practices the damage may be much reduced. Deforestation for large scale agricultural production of arable crops or the provision of grazing land for animals has often been found to be unsustainable without careful scientifically based soil management practices. Small scale forest clearance by resource poor farmers is also a widespread cause of deforestation. The World Bank (1991) has estimated that approximately 60% of the current tropical deforestation (which is variously estimated as being between 10 and 15 million hectares annually) is carried out by small scale farmers for agricultural settlement, often using traditional slash and burn methods. Following the often rapid decline in the productivity in the cleared land, the farmers will abandon it and move on to clear fresh land. This decline in productivity and abandonment is not unique to small scale cultivators. Where land has been extensively cleared for grazing, the productivity of these pastures has been initially relatively poor and has often declined from this poor initial condition (cf. Hecht (1984) who reports poor sustainability of cleared lands for grazing in Amazonia, Brazil). When tropical rainforests have been cleared to be replaced by some form of plantation or mixed tree-crop system the probability of sustained productivity has often been increased. How is it that the lush productive natural rainforest ecosystem when disturbed and deforested is only infrequently replaced by productive managed ecosystems? This paper will attempt to address this paradox by examining the nature and role of the soil component in the

49

B.K. Maloney (ed.), Human Activities and the Tropical Rainforest, 49-64.
© 1998 *Kluwer Academic Publishers. Printed in the Netherlands.*

natural rainforest ecosystem and the impact of deforestation on the soil and its productive capacity.

2.   Soils of the tropics

The term 'tropical soils' is of little use other than to identify the latitudes in which the soils occur, i.e. between 23° North and South. There is no uniquely tropical soil, just as there is no uniquely temperate soil, and there is probably as much variability of soils within the tropics as there is in other regions. It is generally accepted, however, that highly weathered soils are more common in the tropics.

Within the lowland tropics, where the majority of tropical rainforests are found, the major soils (using the classification of Soil Taxonomy (Soil Survey Staff, 1975) are Oxisols (FAO/UNESCO 1984 equivalents; Ferralsols), Ultisols (FAO/UNESCO equivalents; Acrisols and Nitisols) and Alfisols (FAO/UNESCO equivalent; Lixisols and Nitisols (previously Luvisols)).   These three orders are estimated to occupy approximately 50% of the land area of the tropics. Table 1 lists the eleven Orders of Soil Taxonomy and their approximate areal extent and percentages within the tropics.

TABLE 1.   Areal extent of Soil Orders within the tropics

Soil Order	Area ('000km^2)	%
Histosols	286	0.5
Andisols	1683	3.4
Spodosols	40	0.1
Oxisols	115512	23.1
Vertisols	2189	4.4
Aridisols	9117	18.4
Ultisols	9018	18.2
Mollisols	234	0.5
Alfisols	6411	12.9
Inceptisols	4565	9.2
Entisols	3256	6.6
Misc. Land	1358	2.7
Total	49669	100

The Oxisols and Ultisols include soils which are considered to be mildly to severely infertile (in terms of agricultural potential), while Alfisols are considered moderately to highly fertile. Oxisols are the only soil order that occur solely in tropical regions (although there may be relict oxic features in extra-tropical soils).   These soils are the result of strong weathering and desilication in humid, freely draining environments and are characterised by an Oxic subsurface horizon in which the clay fraction is dominated by kaolinite and/or gibbsite.   Characteristically these soils are acidic (pH < 5.0) throughout the soil profile with low cation exchange capacity (< 10 cmol$_c$ kg^{-1} clay).   Much of the retention of cations is due to organic matter found largely in the surface horizons.   Whilst Al^{3+} is the dominant cation on the exchange complex, the absolute amount is small compared to that found in Ultisols because the cation exchange capacity is so low.

Alfisols are characterised by a subsurface horizon of increased clay content (an argillic horizon) due to the accumulation of clay removed from upper soil horizons, translocated down through the profile, and deposited in a lower horizon. In addition this subsurface layer will have a percentage base saturation in excess of 35%, or a profile in which the base status shows an increase with depth. When found in the tropics, these soils tend to have a clay fraction dominated by low activity clays, chiefly kaolinite, but may include 2:1 clays. Because of this dominantly low activity clay, the cation exchange capacity may be relatively low (< 24 cmol$_c$ kg-1 clay), consequently the percentage base saturation should be interpreted with caution - the high base saturation of these soils does not necessarily mean that they are 'rich' in bases. These soils do frequently contain small amounts of micas and other weatherable minerals and are often considered to be the relatively high base status soils of the lowland tropics (cf. Sanchez and Buol 1975). Two major suborders are found in the tropics:

> Udalfs which only have short periods when the soil is dry;
> Ustalfs which have more than 90 cumulative days when the soil is dry.

The Ultisols also have an argillic horizon, but in contrast to Alfisols these soils have a low supply of bases, particularly with depth, in the profile. The clay fraction of these soils is dominated by kaolinite with some 2:1 minerals present. The soils have a low cation exchange capacity (< 24 cmol$_c$ kg^{-1} clay) similar to that of Alfisols, but often have a low pH (< 5.5), which frequently results in exchangeable aluminium dominating the exchange complex, particularly below the surface horizons. These levels of exchangeable aluminium may impose severe agronomic restrictions on the use of this soil.

It is therefore obvious that in many tropical forested systems the use of the land following deforestation will be constrained to a very large degree by the nature of the soils. The soils found most extensively within this region which are likely to afford the greatest constraints to agricultural production on deforested lands are Oxisols and Ultisols. Nortcliff (1989), reviewing the soil and soil related constraints to development in the Amazonian region of South America, identified broad categories of soil based constraints to agricultural development for these soils summarised in Table 2.

TABLE 2.. Major soil based constraints to agricultural development in deforested tropical lowlands in Amazonia.

Oxisols	Ultisols
1. low plant nutrient stores in the soil	1. low plant nutrient stores in the soil
2. low (often exceptionally low) cation exchange capacity	2. high levels of exchangeable aluminium
3. weak retention of bases	3. strong nitrogen losses in the surface layers through leaching
4. strong fixation or deficiency of P	4. low structural strength in surface layers (particularly as soil organic matter levels decline)
5. strong nitrogen losses (particularly in the surface layers) through leaching	5. decrease in permeability of the argillic B horizon
6. strong acidity	
7. high levels of exchangeable aluminium	
8. very low calcium content	
9. restricted rooting volume as a result of aluminium toxicity may result in moisture stress.	

3.   Tree-Soil Interactions

There has been very limited investigation into the tree-soil interactions in the context of natural tropical rainforest systems but they have been more intensively studied in the context of agroforestry. Sanchez (1995) has recently reviewed the scientific background to the soil-tree interactions in agroforestry systems and Lal (1991) considered aspects of soil: tree interactions in the context of sustainable management for soils in the tropics. Whilst agroforestry practices are not restricted to tropical and sub-tropical regions, it is in these regions that their use is most widespread, often because the land users anticipate some benefit from the inclusion of trees along with the arable crops in the farming system. Agroforestry systems have frequently become a key element in attempts to produce sustainable land use practices, often following the clearance of natural rainforest, in particular these practices have often been considered important on poor, low fertility soils such as Oxisols and Ultisols. The claimed benefits of mixing trees and crops, either spatially in, for example, some form of hedgerow intercropping system, or sequentially in a fallow system, serve as a basis for the consideration of what benefits might be removed when tropical forested lands are deforested. Some of the benefits claimed for trees in agroforestry systems are listed below together with their relevance to tropical forested systems.

1.   Trees as components of agroforestry systems afford protection to the soil
        surface by intercepting rainfall and possibly interrupting surface runoff if this
        occurs (cf. Alegre and Fernandes 1991; Banda *et al.* 1994; Kiepe and Rao 1994;
        and Young 1989). Given the often multi-layered canopy of tropical rainforests
        this beneficial effect is likely to be much enhanced in natural forests.

2.   Trees in agroforestry systems may maintain soil organic matter at satisfactory
        levels to preserve overall soil fertility and productivity. Whilst there is very
        little evidence to support this claim in agroforestry systems (Kang *et al.* 1990)
        provide evidence for beneficial effects on sandy soils), there is a substantial
        body of evidence for rapid decline in soil organic matter levels following
        deforestation (cf. Nye and Greenland, 1960)

3.   It is often claimed that agroforestry systems improve soil physical properties in
        comparison to systems which do not incorporate trees (Van Noordwijk *et al.*
        1992). There is considerable evidence that soil physical properties are
        maintained at a relatively beneficial level due to large organic matter inputs
        from continuous litter fall in tropical rainforests (Ewel 1976) and that these
        physical properties rapidly decline following clearance of rainforest (cf.
        Nortcliff and Dias, 1985).

4.   Nitrogen fixing trees can substantially augment nitrogen inputs into the
        system. Ladha *et al.* (1993) have provided clear evidence for this in
        agroforestry systems but there is little evidence of the importance of nitrogen
        fixing trees in the diverse rainforest system.

5.   Trees in agroforestry systems provide for deep nutrient capture from within the
        soil : regolith system. Whilst this has been shown to occur only in selected
        environments (Hartemink *et al.* 1996) for agroforestry systems, it has been

assumed to be a key element in the efficiency of the soil: tree nutrient cycling system in natural rainforest systems.

6.   Agroforestry systems can lead to a more closed nutrient cycle and to more efficient use of nutrients and less leaching losses. This has long been the one of the underpinning hypotheses for the sustainability of tropic rainforest systems particularly on poor soils (cf. Jordan 1985).

These 'hypotheses' of the beneficial effects of trees in agroforestry systems are in broad terms at least applicable to soil: tree interactions in the natural rainforest systems. Removal of the trees through deforestation destroys these interactions and the consequences may be serious for both the soil and the environmental system as a whole

4.   Soils in Forested Ecosystems

Tropical rainforests thrive on both fertile and infertile soils, however the vast majority are found on the highly weathered soils reviewed above. Bruijnzeel (1991) suggested that while the classical view that tropical rainforests are exclusively found on nutrient poor soils may be an over generalisation, almost 80% of soils underlying forests in the moist lowland tropics are of moderate to low fertility. In their natural undisturbed condition these soils sustain a productive ecosystem. Richards (1952) suggested that nutrients may limit the productivity of tropical forested ecosystems, but it is clear that while the productivity of the systems may be limited, it is still high. Rainforests are characterised by high biomass and productivity. It has long been proposed that tropical rainforests retain the bulk of their nutrients in the forest biomass as a conservation mechanism to prevent loss (Anderson and Spencer 1991). In natural ecosystems these conditions are maintained by exceedingly efficient nutrient cycling. Analysis of the available data suggests that such a generalisation is not sustainable. Table 3 presents data for the proportions of the five major nutrients held in the soils of five tropical rainforest systems. These data clearly illustrate that the proportions held in the soil are very variable, both with respect to the particular forested ecosystem and the nutrient considered. The proportion of N held in the soil is reported to be as high as 95.4% in New Guinea rainforests, but as low as 37.6% in the distinctive forests of San Carlos in Venezuela. Similarly the proportion of P found in the soil ranged from 24.8% in New Guinea to 50.6% in Brazil. Wide variations are also reported for Ca and Mg, but the proportion of K held in the soil is low.

TABLE 3. Proportion of nutrients in the soil (excluding roots and litter) system of tropical rainforest ecosystems, expressed as a percentage of total ecosystem nutrients

	N	P	K	Ca	Mg
Venezuela[1]	37.6	33.8	10.0	26.2	7.8
Brazil[2]	56.5	50.6	10.3	0.0	5.8
Panama[3]	n.a.	11.7	11.3	84.3	83.7
Ghana[4]	68.5	7.7	n.a.	n.a.	n.a.
New Guinea[5]	95.4	24.8	31.7	68.2	71.9

Herrera (1979),[1] Klinge (1976),[2] Golley et al. (1975),[3] Greenland and Kowal (1960),[4] Edwards (1982)[5]

The total amounts of nutrients accumulated in the forest biomass of a given forest ecosystem will depend upon the inherent fertility of the soil (Furch and Klinge 1989). Forests growing on less fertile soils are less productive and as a result cycle smaller amounts per year. Carbon and nutrients are transferred to the soil by litterfall and below ground root senescence and exudation. Litterfall additions are reported to range from 10.6 t ha^{-1} a^{-1} to 5 t ha^{-1} a^{-1} (Furch and Klinge 1989; Vitousek and Sandford 1986). There are very few estimates of below ground inputs, Vitousek and Sandford (1986) suggested a maximum of 16 t ha^{-1} a^{-1}.

TABLE 4.   Nutrient pools in mature tropical rainforests

	Above ground biomass kg ha^{-1}					Root biomass kg ha^{-1}				
	N	P	K	Ca	Mg	N	P	K	Ca	Mg
Manaus, Brazil[1]	1428	59	434	424	202	581	7	52	83	55
Kade, Ghana[2]	1568	106	774	1959	289	214	11	88	146	44
El Verde, Puerto Rico[3]	1021	59	926	1129	253	27	6	21	29	8
Bajo Calima, Colombia	1210	36	505	337	386	526	1	128	144	63

Klinge et al. (1975)[1], Greenland and Kowal (1960),[2]Golley et al. (1975),[3]Rodriguez (1989).[4]

There is considerable variability in the nature of the various components of the forest systems with respect to the pools of plant nutrients. This variability is evident both with respect to the overall magnitude of the individual nutrients in contrasting forests and the apportionment to above and below ground components. Table 4 presents data for the mass of nutrients stored in four mature tropical rainforests in both the above and below ground pools.

While there will be major contrasts in the patterns of cycling of individual nutrients, it has been proposed that there are two broad nutrient cycling mechanisms when viewed from the perspective of the wider ecosystem. The relative importance of these processes is dependent upon the nature of the soil:

1.   Internal or closed cycling involves the rapid cycling of nutrients which arrive at the forest floor in litter, generally facilitated by a concentration of fine roots near the soil surface. The vast bulk of the nutrients are held in the biomass, this may be particularly so for N and P (Scott 1986; Jordan 1985). It is suggested that rainforests in the Amazon and Congo basins depend upon internal cycling to satisfy their nutritional requirements.

2.   External or open cycling involves the cycling of nutrients from weathering in the subsoil and saprolite layers in addition to the internal cycling. The nutrient provision under these conditions is closely related to the composition of the parent rock. Close correlations between 'reserve' quantities of K, Ca and Mg in the soil and rock and the vegetation have been shown for dipterocarp forests in Southeast Asia (Baillie and Ashton, 1983). These processes have been

suggested to be significant in rainforests on Alfisols and Andisols of Central America and West Africa and in the extensive forests of Malaysia, Indonesia and New Guinea found over shallow less weathered soils (Lamb 1980; Bruinjzeel 1989; Burnham 1989).

When considering the potential consequences of deforestation it is important to differentiate between tropical rainforest systems which are sustained by weathering release of nutrients in 'open cycling' and those where 'closed or nearly closed cycles' predominate (Baillie 1989; Burnham 1989).

5. Clearance using 'slash and burn' techniques

Where a substantial proportion of the nutrients are held in the biomass with the soil input as a relatively minor contributor, the destruction of the rainforest through felling will drastically change the nature and magnitude of the nutrient pools. If the biomass is exported in the form of timber and other wood products there will be a substantial export of nutrients from the system. If the soils are poor and infertile the loss of in excess of 80% of the total nutrient pool may occur if the biomass is exported. Frequently, however, where the land is deforested to provide land for cultivation the trees and associated forest debris are left on site and, at least in part, burnt. The process of 'slash and burn' is for many the only practical and economic option to facilitate the provision of land for agricultural activity. The changes in soil temperature that result from burning the vegetation depends upon the duration of the fire and its intensity. The composition of the vegetation and the combustion rate will determine the intensity of the fire. Sanchez (1976) reported temperatures of 450°C to 650°C at 2 cm above the soil surface, which decreased in the top 5 cm at a rate of 100°C per cm. These high temperatures may locally activate the destruction of organic substances in the soil and have a direct detrimental effect on the soil microflora and fauna, but less readily available mineral nutrients may become more available. During this burning process there may be a loss through volatilisation, but the bulk of the nutrients will be retained in the system, in the ash and unburnt organic materials. The ash will contain nutrients in particularly soluble forms. If these readily soluble nutrients are not rapidly 'captured', preferably in a form which does not render them unavailable in the long term, they may be lost through leaching and in run off and possibly as aerosols. There are two properties of tropical soils which may assist in reducing the losses from the system in soluble forms. The first is the presence of positive charges in subsurface layers of many Oxisols and Ultisols, which retard the loss of anions such as nitrate and sulphate (Wong et al. 1987). This retardation process extends the period of availability for uptake by plants if there is a rooting system extending deep in to the soil. A second property of some soils within the tropics which may assist in the reduction of losses from leaching is the bi-phasic nature of the pore system. A number of workers (cf. Grimme and Juo 1985; Pleysier and Juo 1981; Nortcliff and Thornes 1989) have observed a distinct two phase pore systems and associated differences in the movement of water through well weathered highly permeable tropical soils. These studies identified rapid flow through the macropore phase, with rainfall events filling and emptying of this phase. In contrast, in the second phase the micropores appeared to behave almost independently of the rainfall inputs, filling and emptying of water over longer time spans. These authors hypothesised that nutrients were held in solution available, for plant uptake, in the micropore system, whereas in contrast the macropore flows bypassed the bulk of the soil and contributed little to nutrient loss.

An important benefit of slash and burn deforestation is the rapid release of mineral nutrients such as Ca, Mg, K and P from the ash to the soil. The benefits may be short lived if the soil system is unable to store this large addition of soluble nutrients against losses through runoff, leaching and erosion during the crop establishment and cropping phase. The degree to which these nutrients are retained varies considerably. Jordan (1985) reporting results from the Amazon Basin observed that nearly all the mineral nutrients from the ash were initially retained in the system, but in the succeeding three years there was a substantial loss through leaching. In contrast Tulaphitak *et al.* (1985) reported results from sites on steeplands in Thailand where more than 50% of Ca, Mg, K and P released from the biomass during clearance was lost through erosion and runoff at the onset of the rainy season immediately after clearance.

In slash and burn systems, the decline of the ash enriched nutrient pool of the soil during the cropping phase may be accelerated under the following conditions:

1.   lack of continuous ground cover, increasing the susceptibility of the soil to erosion and runoff losses during rainstorms.

2.   increased frequency of clearance and cultivation leading to a gradual destruction of the macropore systems with a resultant decrease in rapid bypass flow, increasing runoff and leaching losses.

3.   burning and destruction lead to gradual destruction of the natural forest root mat, decomposition of the humified organic matter and a reduction in the microbiological activity in the soil.

The physical properties of soils following shifting cultivation practices of 'slash and burn' tropical deforestation are often relatively little altered (cf. Dias and Nortcliff 1985), but once exposed to the direct impact of rainfall following the removal of the 'protective cover' of the forest, the soils may be subject to rapid physical deterioration and possibly the onset of soil erosion. Under similar conditions of rainfall, slope, soil cover and soil particle size distribution, Oxisols are generally considered to be less vulnerable to physical deterioration and the onset of soil erosion than Ultisols. Oxisols with their generally more stable aggregation have higher infiltration rates, even when the soils have a high proportion of clay because of the presence of the well connected macropore system. If following deforestation the soil surface is left bare the strong aggregation assists to resist erosion, but some erosion may take place depending upon the nature of the rainfall characteristics. The soil aggregation present in the Ultisols is characteristically weaker than in the Oxisols, consequently soil aggregates at the surface of Ultisols are more readily broken down by raindrop impact. In recently deforested lands the soil erosion losses tend to be greater on Ultisols than on Oxisols. Taking the Amazon basin as an example, Cochrane and Sanchez (1982) suggested that only 6% and 10% of the soils in the tropical rainforest and semi-evergreen rainforest respectively are highly erodible. Smyth and Cassel (1995) have considered these estimates to be on the low side for the Amazon Basin, and given the relatively low relief of much of the Amazon basin, may be gross underestimates for the potential erosion in tropical deforested areas with more relief.

6.   Clearance using mechanical methods

The alternative to 'slash and burn' clearance where trees are manually felled is to use some form of mechanical means either to fell the trees or to move the trees and associated debris in a post felling clearing phase. It has been widely reported that the use of heavy machinery during deforestation often results in substantial alterations to the soil physical conditions, alterations which may have significant impacts on the post clearance use of the soil (cf. Alegre *et al.* 1986; Lal *et al.* 1975; Lal and Cummings 1979; Nortcliff and Dias 1988). Soil compaction is often a significant consequence of these broad group of mechanical clearance methods. This results in changes in the structural arrangement of soil particles and pores. Consequences of this compaction are often changes in the infiltration characteristics of the soil (often a reduction both in the infiltration rate and infiltration capacity), poor soil aeration and mechanical impedance for growing roots and seedling emergence.

Alegre *et al.* (1986) and Alegre and Cassel (1986) report comparisons of the effects on soil physical properties of a range of mechanised forest clearance activities with traditional slash and burn clearance on an Ultisol at Yurimaguas in the Peruvian Amazon. They showed significant detrimental changes in soil bulk density, infiltration rate and aggregate stability association with mechanical clearance methods in comparison to slash and burn clearance. Dias and Nortcliff (1985) found similar stark contrasts when comparing bulldozer clearance with traditional slash and burn clearance practices on an Oxisol in the central Amazon basin in Brazil. Table 5 shows the mean dry bulk densities recorded to 20 cm in the soil profile of an Oxisol near Manaus cleared by traditional and mechanical methods.

Nicholaides *et al.* (1983) present data showing marked reductions in infiltration rate under natural forest, 15 year fallow, and at two stages following clearance using traditional and bulldozer methods. These results illustrate the marked impact on soil physical properties of any disruption to the forest system, whether by traditional or mechanical methods. They also show that the system is capable of recovery when damaged during mechanical clearance even under a cropping regime, but that the recovery is slow. It should be noted that the infiltration rates are still high and only in the case of the bulldozed site after one year of cropping are rainfall intensities likely to be in excess of infiltration rates except under extreme conditions.

In addition to the direct effect of mechanical methods of clearance on soil physical properties, the reduction in structural stability due to increased mineralisation of soil organic matter has been cited by Alegre and Cassel as a possible consequence of mechanical clearance methods. Nye and Greenland (1960) also noted increases in mineralisation of soil organic matter following clearance. Martin *et al.* (1991) investigated the linkages between changes in soil organic matter pools and soil structural properties. They suggested that following deforestation and tillage in the Amazon basin the decreasing content and changes in the forms of soil organic matter resulted in a progressive destruction of clay-organic complexes in the surface soil, fine particles were dispersed due to this and migrated down the profile causing obstruction of macropores and an increase in soil bulk density. A a rapid decrease in crop production and abandonment of the site after two years was associated with this decline in soil conditions.

TABLE 5. Mean dry bulk density under virgin forest, slash and burn clearance and bulldozer clearance with depth.

Depth (cm)	Mean dry bulk density (standard error)		
	virgin forest	slash and burn	bulldozed
0-5	0.79 (0.02)	0.84 (0.02)	1.31 (0.01)
5-10	1.04 (0.01)	1.06 (0.01)	1.25 (0.02)
10-15	1.12 (0.02)	1.13 (0.01)	1.17 (0.02)
15-20	1.11 (0.02)	1.12 (0.02)	1.18 (0.02)

TABLE 6. Effects of land clearing methods on infiltration rates in Ultisols at Yurimaguas, Peru.

	Infiltration rate $(cm\ h^{-1})$
Virgin Forest	117.0
15 year forest fallow	81.0
Slash and burn (after 1 year cropping)	10.0
Bulldozed (after 1 year cropping)	0.5
Slash and burn (after 6 years cropping)	10.0
Bulldozed (after 6 years cropping)	4.1

The physical changes in soil conditions following clearance may be reflected in crop yields. Seubert *et al.* (1977) present data for the response of maize (one harvest), soybeans and cassava (means of two harvests) planted on land cleared by traditional and mechanical methods to three fertiliser management practices; 0 - no fertiliser or lime; NPK - 50 kg ha^{-1} N, 172 kg ha^{-1} P, 40 kg ha^{-1} K; NPKL - 50 kg ha^{-1} N, 172 kg ha^{-1} P, 40 kg ha^{-1} K, 4 t ha^{-1} lime. These results are presented in Table 7, and illustrate the varying influence of the changed soil physical conditions on yield of different crops and the degree to which these yield differences may be 'overcome' by addition of plant nutrients in the form of fertilisers and by relieving the soil acidity constraint by the addition of lime.

TABLE 7. Yields of maize, soybean and cassava in response to additions of fertiliser and lime and land clearance activity (t ha^{-1}).

Crop	Fertiliser Treatment	Slash and Burn Clearance	Bulldozer Clearance
Maize(grain)	0	0.1	0.0
	NPK	0.44	0.04
	NPKL	3.11	2.36
Soybeans (grain)	0	0.70	0.15
	NPK	0.95	0.30
	NPKL	2.65	1.80
Cassava (roots)	0	15.4	6.4
	NPK	18.9	14.9
	NPKL	25.6	24.9

7. Deforestation and Soil Erosion

A widely seen consequence of deforestation is soil erosion, particularly on steeply sloping land. Lal (1990) in reviewing soil erosion in tropical environments noted that under undisturbed natural forests the predominant forms of soil erosion are: splash erosion, sheet flow and soil creep. Lal observed that when forested systems are disrupted more extensive erosion in the form of rills and gullies is frequently seen. Where other factors are the same Lal suggests a broad trend in erosion hazard: bare fallow > arable land > perennial crops > grass cover > natural forest vegetation. Francis and Thornes (1990) have also demonstrated for a wide range of environments that both runoff and sediment yield fall exponentially as the percentage vegetation cover increases. Under natural rainforest because of the multi-layered canopy the percentage vegetation cover may approach 100%, consequently any disruption of the forest system is likely to produce a reduction in the cover and substantially increase the potential for erosion. These relationships will become more critical where the forest is on steeplands, for example, Agustin and Nortcliff (1994) report very high erosion rates on recently cleared forested lands on steeplands in northern Luzon, in the Philippines.

Nortcliff *et al.* (1990) and Ross *et al.* (1990) presented results of an experimental investigation into the effects of forest clearance and management on runoff and erosion in Roraima, northern Brazil. This study compared three treatments (natural forest, partial clearance and total clearance) at three positions down a catenary sequence under seasonal rainforest. The natural forest sites were left minimally disturbed and the two sets of clearance sites differed in degree of clearance. In the 'Total Clearance' sites the vegetation (both tree and shrub layers) and the surface litter were removed from the site. For the 'Partial Clearance' site only the tree layer was removed, the shrub and litter layers were retained intact. The sites were monitored during 1987 immediately following the clearance. The results (Table 8) show the marked contrast between the totally cleared site on the one hand and the undisturbed forest and partially cleared site on the other.

TABLE 8. Effects of clearance on runoff and sediment yield for three treatments (27/05/87-
          22/11/87) in Roraima, northern Brazil.

Runoff (litres m^{-2})	Top slope	Mid slope	Bottom slope	
Undisturbed Forest	76.31	98.36	128.25	
Partially Cleared Forest	89.80	131.05	65.00	
Totally Cleared Forest	80.20	361.10	255.20	
Sediment Yield (kg m^{-2})				
Undisturbed Forest	1.44	1.28	2.42	
Partially Cleared Forest	2.72	5.17	2.37	
Totally Cleared Forest	3.93	52.40	37.50	

This illustrates the complexity of the system and, perhaps, suggests that a key element in managing the system to minimise erosion following clearance may be to develop practices that maintain the near surface shrub layer and the 'protective cover' of litter at the soil surface. This might possibly be achieved by a programme of phased clearance with an initial clearance of the tree component followed by an interim period when the natural shrub and litter layer is managed to allow replacement by the 'productive crop'. These results are supported by the suggestions of Wiersum (1985) and Brandt (1986) that during deforestation it is the removal of the understorey and the surface litter that are more important than removal of the tree canopy in determining the vulnerability of the soil to erosion. Both Wiersum and Brandt noted a marked increase in rainfall energy at the soil surface when natural forest was removed, in particular they noted high rainfall energies where the replacement for the natural multi-storied rainforest was by a single storey high canopy plantation.

It is obvious therefore that soil erosion is not an inevitable consequence of deforestation, but when the trees are removed the soil is exceptionally vulnerable to erosion, particularly given the high energy rainfall regimes characteristic of the humid tropics. Soil erosion may be avoided if steps are taken to manage the deforested system to minimise the potential for erosion, these may involve deforestation as a phased process, with the initial removal of trees followed by a period when the lower layers in the forested system and the ground litter are left relatively undisturbed until the replacement vegetative cover offers a degree of ground protection, failure to do this may almost inevitably result in the rapid onset of severe erosion.

8. Conclusion

Tropical rainforests, widely underlain by soils of moderate to low fertility have adapted to this scarcity of nutrients in the soil by developing a number of conserving mechanisms, perhaps the most important of which is the concentration of fine roots at or near the soil surface to 'capture' nutrients released from the decomposition of litterfall. Upon large scale mechanised clearance and burning of the forest, the root mat becomes disrupted, potentially causing substantial losses of much of the nutrient capital in the system through leaching, particulate loss in erosion and volatilisation, if the clearance is not managed to prevent these losses. In addition if the timber of the forest is harvested there will be a further substantial export of nutrients. Associated with the loss of nutrients there is frequently a rapid deterioration in soil physical properties, a deterioration which may be enhanced if mechanical clearance methods are used.

It is also the case that rainforests play a significant role in the conservation of natural resources, stability of climate and control of the quality of the environment. In the context of soils, rainforests provide maximum protection against erosion and therefore conserve the soil, preventing damage to the soil and the impacts on other components of the system. Given the close but often fragile association between the soil and the forest and the key role the rainforests play with respect to other environmental components why does the rate of tropical deforestation continue to increase? If rainforests are considered economically, they provide very few commodities per unit area that would justify their acceptance as economically valuable assets. The consequence has been that there has been considerable activity to fell the rainforest to 'release' other resources. One of these resources is the soil, but as is shown above this resource exists together in close association with the rainforest and disruption of the rainforest through deforestation may result in disruption and possible destruction of the soil resource.

Van Wambeke (1992) also reviewed this situation and suggested that if, viewed in the light of its economic potential, the rainforest might be considered no more than a mass of weeds, albeit enormous weeds that have to be removed before any resourceful land use system can be implemented. Such destruction may have potentially devastating effects to the forest, the soil and other components of the system, but if the aim is to increase the economic output then destruction is necessary. The detrimental consequences of the deforestation may be reduced if careful attention is given to the criteria for selection of the sites to be deforested. Selection criteria might include soils, topography and the requirements of the planned land utilisation. In economic terms the nature of the forest is probably irrelevant.

If this is to be the strategy for deforestation, how does our knowledge of the interactions between the soil and rainforest assist in the decision making? Are we able to identify which conversions of forested lands are likely to lead to unsustainable land use systems which will lead to land degradation and require further deforestation to satisfy the demand for land? In contrast are we able to identify more sustainable land use systems which with careful management will maintain output and slow down the rate of deforestation and conversion of forested lands to other uses? To prevent deforestation, which inevitably produces wholesale degradation of the rainforest, soil and other components of the environmental system, the close association which exists between the soil and the rainforest must be emphasised and the nature of this association needs to be fully understood. Deforestation which does not take account of these associations and the nature of the processes involved has a high probability of producing the environmental devastation which is so widely seen throughout the tropics.

References

Agustin, E.O. and Nortcliff, S. (1994). Agroforestry practices to control runoff and erosion on steeplands in Ilos Norte, Philippines, in J.K. Syers and D.L. Rimmer (eds) *Soil Science and Sustainable Land Management in the Tropics*, CABI, Wallingford, pp. 59-72.

Anderson, J.M. and Spencer, T. (1991). Carbon, nutrient and water balance of tropical rain forest ecosystems subject to disturbance: management implications and research proposals. *MAB Digest* **7**, UNESCO, Paris.

Alegre, J.C. and Fernandes, E.C.M. (1991). Runoff and erosion losses under forest, low input agriculture and alley cropping on slopes, in *Tropical Soils Technical Report, 1988-89*, North Carolina State University, Raleigh, NC. pp. 227-228.

Baillie, I.C. (1989). Soil characteristics and classification in relation to the mineral nutrition of tropical wooded ecosystems, in J.C. Proctor (ed) *Mineral Nutrients in Tropical Forest and Savanna Ecosystems*, pp. 15-26, Blackwell Science Publications, Oxford.

Baillie, I.C. and Ashton, P.S. (1983). Some aspects of the nutrient cycle in mixed dipterocarp forest in Sarawak, Malaysia, in S.C. Sutton, T.C. Whitmore and A.C. Chadwick (eds) *The Tropical rainforest: Ecology and Management*, Blackwell Scientific, Oxford, pp. 347-376.

Banda, A.Z., Maghembe, J.A., Ngugi , D.N. and Chome, V.A. (1994). Effect of intercropping and closely spaced leucaena hedgerows on soil conservation and maize yield on a steep slop at Ntcheu, Malawi, *Agroforestry Systems*, **27**, 17-22.

Brandt, J. (1986). Transformations of the kinetic energy of rainfall with variable tree canopies. Unpublished PhD thesis, University of London, England.

Bruijnzeel, L.A. (1989). Nutrient cycling in moist tropical forests: the hydrological framework, in J.C. Proctor (ed) *Mineral Nutrients in Tropical Forest and Savanna Ecosystems*, Blackwell Science Publications, Oxford, pp. 383-413.

Bruijnzeel, L.A. (1991). Nutrient input-output budgets of tropical forest systems: a review, *Journal Tropical Ecology*, **7**, 1-24

Burnham, C.P. (1989). Pedological processes and nutrient supply from parent materials in tropical soils, in J.C. Proctor (ed) *Mineral Nutrients in Tropical Forest and Savanna Ecosystems*, Blackwell Science Publications, Oxford, pp. 27-42.

Cochrane, T.T. and Sanchez, P.A. (1982). Land resources, soils and their management in the Amazon region: a state of knowledge report, in S.B. Hecht (ed) *Amazonia, Agriculture and Land Use Research*, CIAT, Cali, pp. 137-209.

Dias, A.C.P. and Nortcliff, S. (1985). Effect of two land clearing methods on the physical properties of an oxisol in the Brazilian Amazon, *Tropical Agriculture*, **62**, 207-213.

Edwards, P.J. (1982). Studies of mineral cycling in a montane rainforest in New Guinea V. Rates of cycling in throughfall and litterfall, *Journal of Ecology*, **70**, 807-828.

Ewe, J. (1976). Litterfall and decomposition in a tropical forest succession in eastern Guatemala, *Journal Ecology*, **64**, 293-308.

FAO (1983). Keeping the soil alive - soil erosion its causes and cure, *Soil Bulletin 50*, FAO, Rome

FAO/UNESCO (1984). *Soil Map of the World: Legend*, FAO, Rome.

Francis, C.F. and Thornes, J.B. (1990). Runoff hydrographs from three Mediterranean Cover Types, in J.B. Thornes (ed) *Vegetation and Erosion*, John Wiley, Chichester, pp. 363-384.

Furch, K. and Klinge, H. (1989). Chemical relationships between vegetation, soil and water in contrasting inundation areas of Amazonia, in J.C. Proctor (ed) *Mineral Nutrients in Tropical Forest and Savanna Ecosystems*, Blackwell Science Publications, Oxford, pp. 189-204.

Golley, F.B., McGinnis, J.T., Clements, R.G., Child G.I. and Deuver, M.J. (1975). *Mineral Cycling in a Tropical Forest Ecosystem*, University of Georgia Press, Athens, GA.

Greenland, D. J. and Kowal, J.M.L. (1960). Nutrient content of the moist tropical forests of Ghana, *Plant and Soil*, **12**, 154-174.

Grimme, H. and Juo, A.S.R. (1985). Inorganic nitrogen losses through leaching and denitrification in soils of the humid tropics, in B.T. Kang and J. van der Heide (eds) *Nitrogen Management in Farming Systems in the Humid and Subhumid Tropics*, Haren, Netherlands, pp. 57-73.

Hartemink, A.E., Buresh, R.J., Jama B. and Janssen, B.H. (1996). Soil nitrate and water dynamics in sesbania fallows, weed fallows and maize, *Soil Science Society of America Journal*, in press.

Hecht, S.B. (1984). Cattle ranching in Amazonia: political and ecological considerations, in M. Schmink and C.H. Wood (eds) *Frontier Expansion in Amazonia*, University of Florida Press, Gainesville, GLA., pp. 366-398.

Herrera, R. (1979). Nutrient distribution and cycling in an Amazon caatinga forest on spodosols in Southern Venezuela. Unpublished PhD thesis, University of Reading, England.

Jordan, C.F. (1985). *Nutrient Cycling in Tropical Forest Ecosystems*, John Wiley and Sons, New York.

Kang, B.T., Reynolds, L. and Atah-Krah A.N. (1990). Alley farming, *Advances in Agronomy*, **43**, 315-359.

Kiepe, P. and Rao, M.R. (1994). Management of agroforestry for the conservation and utilisation of land and water resources, *Outlook on Agriculture*, **23**, 17-25.

Klinge H. (1976) Bilanzierung von Hauptnahrstoffen in okosystem tropischser Regenwald (Manaus) - vorlaufen Daten, *Biogeographia*, **7**, 59-76.

Klinge, H., Rodriguez, W.A., Brunig, E. and Fittkau, E.J. (1975). Biomass and structure in a central Amazonian rainforest, in F.B. Golley and E. Medina (eds) *Tropical Ecological Systems*, Springer Verlag, New York, pp. 115-122.

Ladha, J.K., Peoples, M.B., Garrity, D.P., Capuno, V.T. and Dart, P.J. (1993). Estimating dinitrogen fixation of hedgerow vegetation using the nitrogen-15 natural abundance method. *Soil Science Society of America Journal*, **57**, 732-737.

Lal, R. (1990). *Soil Erosion in the Tropics*, McGraw Hill, New York.

Lal, R. (1991). Myths and scientific realities of agroforestry as a strategy for sustainable management for oils in the Tropics, *Advances in Soil Science*, **15**, 91-137.

Lal, R. and Cummings, D.J. (1979). Changes in soil and micro-climate by forest removal. *Field Crops Research*, **2**, 91-107.

Lal, R., Kang, B.T., Moorman, F.R., Juo, A.S.R. and Moomaw, J.C. (1975). Soil management problems and possible solutions in western Nigeria, in E. Bornemisza and A. Alvarado (ed) *Soil Management in Tropical America*, North Carolina State University Press, Raleigh, pp. 372-408.

Lamb, D. (1980) Soil nitrogen mineralisation in a secondary rainforest succession, *Oecologia*, **47**, 257-263.

Martin, P.F. da S., Cerri, C.C., Volkoff, B., Andreux, F. and Chauvel, A. (1991). Consequences of clearing and tillage on the soil of a natural Amazonian ecosystem, *Forest Ecology and Management*, **38**, 273-281.

Nicholaides, J.J. III, Sanchez, P.A., Brandy, D.E., Villachia, J.H., Couto, A.J. and Valverde, C.S. (1983). Crop production systems in the Amazon basin, in E.F. Moran (ed) *The Dilemma of Amazonian Development*, Westview Press, Boulder, Colorado, pp. 101-153.

Nortcliff, S. (1989). A review of soil and soil related constraints to development in Amazonia. *Applied Geography*, **9**, 147-160.

Nortcliff, S. and Dias, A.C.P. (1985). The change in soil physical conditions resulting from forest clearance in the humid tropics, *Journal of Biogeography*, **15**, 61-66.

Nortcliff, S. and Thornes, J.B. (1989). Variation in soil nutrients in relation to moisture status in a tropical forest ecosystem, in J.C. Proctor (ed) *Mineral Nutrients in Tropical Forest and Savanna Ecosystems*, Blackwell Science Publications, Oxford, pp. 43-54.

Nortcliff, S. Ross, S.M. and Thornes, J.B. (1990). Soil mosture, runoff and sediment yield from differentially cleared tropical rainforest plots, in J.B. Thornes (ed) *Vegetation and Erosion*, John Wiley, Chichester, pp. 419-436.

Nye, P.H. and Greenland, D.J. (1960). *The soil under shifting cultivation*, Commonwealth Agricultural Bureaux Technical Communication 51, Harpenden, UK.

Pleysier, J.L. and Juo, A.S.R. (1981). leaching of nutrient ions in an Ultisol from the high rainfall tropics through undisturbed soil columns, *Soil Science Society of America Journal*, **45**, 754-760.

Richards, P.W. (1952). *The Tropical Rain Forest*, Cambridge University Press, Cambridge.

Rodriguez, J.L.V.A (1989). Consideraciones sobre la biomasa, composicion quimica y dinamica del bosque pluvial tropical de Colonias Bajas, Bajo Calm, Buenaventura, Colombia, *CONIF Serie documentacion no. 16*, Bogota, DE, Colombia.

Ross, S.M., Thornes, J.B. and Nortcliff, S. (1990). Soil hydrology, nutrient and erosional response to the clearance of terra firme forest, Maraca Island, Roraima, northern Brazil, *Geographical Journal*, **156**, 267-282.

Sanchez, P.A. and Buol, S.W. (1975). Soils of the Tropics and the World Food Crisis, *Science*, **180**, 598-603.

Sanchez, P.A. (1976). *Properties and Management of Soils in the Tropics*, John Wiley and Sons, New York.

Sanchez, P.A. (1995). Science in Agroforestry, *Agroforestry Systems*, **30**, 5-55.

Scott, G.A.J. (1986). Shifting cultivation where land is limited, in C.F. Jordan (ed) *Sustainable Agriculture and the Environment in the Humid Tropics*, NRC, National Academic press, Washington, DC., pp. 263-351.

Seubert, C.E., Sanchez, P.A. and Valverde, C.S. (1977). Effects of land clearing methods on soil properties of an Ultisol and crop performance in the Amazon jungle of Peru, *Tropical Agriculture*, **54**, 307-321.

Smyth, T.T. and Cassel, D.K. (1995). Long term soil management research in Amazonia, *Advances in Soil Science*, **19**, 13-60.

Soil Survey Staff (1975). *Soil Taxonomy. A Basic System of Soil Classification for Making and Interpreting Soil Surveys*. Agriculture Handbook No. 436, Soil Conservation Service, Washington, DC.

Tulaphitak, T., Puirinta, C. and Kyuma, K. (1985). Changes in soil fertility and soil tilth under shifting cultivation 2. Changes in Nutrient status, *Plant and Soil*, **31**, 239-249.

Van Noordwijk, M., Hairiah, K., Sitomul, S.M. and Syekhfani, M.S. (1992). Rotational hedgerow intercropping + *Peltophorum pterocarpum* - new hope for weed infested soils, *Agroforestry Today*, **4**, 4-6.

Van Wambeke, A. (1992). *Soils of the Tropics: Properties and Appraisel*, McGraw Hill, London.

Vitousek, P.M. and Sandford, R.L. (1986). Nitrogen cycling in moist tropical forests, *Annual Review of Ecological Systematics*, **17**, 137-167.

Wiersum, K.K. (1985). Effect of various vegetation layers of a *Acacia auriculiformis* forest plantation on surface erosion in Java, Indonesia, in S.A. El-Swaify, W.C. Moldenhauer and A. Lo (eds) *Soil Erosion and Conservation*, Soil Conservation Society of America, Arkeny, pp. 77-89.

Wong, M.T.F., Wild, A. and Juo, A.S.R. (1987). Retarded leaching of nitrate measured on monolith lysimeters in south east Nigeria, *Journal of Soil Science*, **38**, 511-518.

World Bank (1991). *The Forest sector: A World Bank Policy Paper*, World Bank, Washington, DC.

Young, A. (1989). *Agroforestry for Soil Conservation*, CABI, Wallingford, UK.

Dr. Stephen Nortcliff, Department of Soil Science, The University of Reading, Reading RG6 6DW, England

4.   THE LONG-TERM HISTORY OF HUMAN ACTIVITY AND RAINFOREST DEVELOPMENT

Bernard K. Maloney

1.   Introduction

This review is concerned with the long term history of rainforest use and development by people and uses the term rainforest in a wide sense to include lowland seasonal forest, e.g. the monsoon forests of Asia, and montane forests as well as what might be regarded as rainforest proper.

Evidence for long term use of the rainforest comes from archaeology, geomorphology and palaeoecology, but only the most indestructible of organic materials survive under lowland tropical rainforest conditions. The main concern here will be with palaeoecological evidence, especially that from pollen analysis, as this has been best studied. Recent data from the world tropics will be mentioned but inevitably the emphasis will be on information from Southeast Asia and Melanesia, the region with which the writer is most familiar, as the amount of literature now available is too vast to assemble comprehensively and condense concisely into a short book chapter.

Archaeological excavations have or are taking place throughout the world tropics in rainforest areas as defined above, in places as diverse as the Congo-Zaire (Eggert 1992), where hunter-gatherer sites have been found, to the highlands of Papua-New Guinea (Golson 1977; Golson and Hughes 1980; Golson and Gardner 1990; Gorecki 1986; Bayliss-Smith and Golson 1992) from where there is dramatic information on the human use of the land thousands of years ago which reflects remarkable early technological developments, which will be returned to later. Recently, and contentiously it has been claimed that the tropical Americas have been occupied for 50,000 years (Bahn 1993; Guidon et al. 1996; Meltzer et al. 1994; Parenti et al. 1990, 1996) and that fire was used, but (Guidon et al. 1996) in the *caatinga* of south east Brazil, not rainforest proper. The earliest occupation of Papua New Guinea dates to about 40,000 B.P. (Groube et al. 1986).

Archaeobotanists are beginning to study plant macrofossils from rainforest area sites (cf. Hather 1994; Kajale 1994; Thompson 1994) but palaeoecologists largely concern themselves with pollen and, more recently, phytoliths (microscopic plant silica). Increasing amounts of palynological information from the world tropics is being published but much of this is not concerned with the possible impact of people on the environment and agricultural origins. These, in contrast, are the main aims of phytolith research. There are phytolith publications relating to the origins of agriculture in Central and South America and Southeast Asia (Kealhofer 1996; Kealhofer and Piperno 1994; Pearsall and Piperno 1990; Piperno et al. 1985; Piperno et al. 1991).

The lowland rainforest is a difficult environment for pre-agricultural peoples (Bailey et al. 1989; Headland 1987; Headland and Reid 1989; Hart and Hart 1986; Eggert 1992; Bellwood 1990) because the plants and animals which provide the food supply are highly diverse, possible staples like tubers do not grow abundantly in primary forest and trees yielding edible fruits or nuts are dispersed, so sources of plant carbohydrates are scarce. Additionally the foods may require prolonged processing to be made edible. It

65

B.K. Maloney (ed.), Human Activities and the Tropical Rainforest, 65-85.
© *1998 Kluwer Academic Publishers. Printed in the Netherlands.*

has been claimed that potential plant foods (tubers, fruits, nuts) are more readily found at forest edges and in natural clearings than in rainforest proper (Mountain 1991). Indeed, tubers thrive best in monsoon forest areas (Eder 1988). The total food biomass, plant or animal, is low.

Most archaeologists agree that agriculture did not originate in rainforest areas, but was spread there and rainforest peoples may have always depended on the resources of non-rainforest areas, exchanging forest products with cultivators from early times (Possehl and Kennedy 1979; Stiles 1993). However, Endicott and Bellwood (1991) disagree claiming that small nomadic groups of foragers can live off wild resources alone, and have done so in the past. Alternatively they may have been only able to subsist in coastal areas, where there was also access to marine resources, at the margins of rainforests or in natural clearings where there was access to a maximum diversity of plant and animal food sources (cf. Gosden 1993: 131)

2.   Prehistoric use of the rainforest

2.1   INTRODUCTION

It could be argued that the fairly limited rainforest destruction which occurred in prehistory and early historic times before the period of European colonial expansion was beneficial to the development of society as it was may have been accompanied increased leisure to create works of art, time to develop complex religious, legal and political systems, in short, with cultural evolution, as well as a more stable food supply.

Where rainforest destruction occurred in the past it was primarily associated with the expansion of agriculture (taken to include horticulture), perhaps with the extension of grazing land or stimulation of pasture grass growth, and, in the mountains, possibly to assist in the hunting of game (Gorecki 1986; Hope and Tulip 1994). However, positive evidence of what was actually cultivated is not easy to obtain (cf. Flenley 1994; Haberle 1994; Maloney 1994) while research on Africa and the tropical Americas has often been less concerned with identification of pollen from possible cultivars than that from Asia, Melanesia and Polynesia, but prehistoric forest clearance is claimed for many areas, as will be discussed later. Plant macrofossils from archaeological sites give much more positive information about which taxa have been cultivated but there are still some identification difficulties (cf. Thompson 1994; Maloney 1993). Again, there is less information from Africa and the American rainforest (as opposed to coastal desert) areas than elsewhere.

Where forest clearance has taken place recently or is occurring today it is usually associated with exploitative destruction for timber, extension of agriculture (e.g. Indonesian transmigration (Conway and McCauley 1983; Harrison 1977; Secrett 1986; Whitten 1987)), extension of grazing land (especially in the Brazilian Amazon, cf Nepstad *et al.* 1994), mining of materials which lie beneath forested land (for instance in Irian Jaya (Suter 1982) and Papua-New Guinea (Hyndman 1988)) or a direct or indirect result of warfare (U.S. aerial bombardment during the Vietnam war and environmental destruction because of massive displacement of refugees from war-torn Rwanda to Zaire are two examples). Where trees are replaced they often comprise alien taxa such as species of *Eucalyptus,* (cf. Barrow this book): the rich species diversity of the rainforest is frequently replaced by grassland, weedy regrowth scrub and monoculture of introduced, fast growing timber trees. Some forest animals remain (Wilson and Johns 1982) but diversities and densities decline. Restitution of tropical rainforest or even dry forest is not easy (Kartawinata *et al.* 1981; Chefas 1986; Holden 1986) but (Wilson and Johns 1982) animal recolonisation of old logged over forest can be successful if hunting levels are low and there are adjacent areas of undisturbed forest. Despite this, Spriggs (1993) suggested that people may have added slightly to forest diversity rather than detracted from it in areas of Melanesia which were occupied at the end of the Pleistocene and in the early Holocene. Generally though there is a great loss of biodiversity which is all the more lamentable in parts of tropical South America and Southeast Asia where the species rich flora has not been well studied. However, it is the past and not the present or the future which is the concern here.

The origins of agriculture and horticulture are impossibly difficult to separate. As elsewhere, agriculture probably began in the tropics with the conservation or deliberate planting of useful plants, followed by introduction of new plants or movement of indigenous plants to new growth environments, accompanied or not accompanied by movements of people. Those plants could have included fruit and nut bearing trees as well as what became agricultural plants. This is a process which must have stimulated plant mutation, if not more significant evolutionary changes.

Agriculture and horticulture may have begun together as part of a shifting cultivation regime. Sedentary agriculture, in contrast, requires more organised, if not greater, manipulation of the environment as well as possession of a staple plant or plants. In both instances wild as well as cultivated forest products were probably used in tandem and wild plants, famine foods, as of neccessity when there was crop failure.

Most palaeoecological evidence for early exploitation of rainforest areas is not from the lowland forest proper but from the mountains, especially, and the coasts where the natural products of more than one environment were almost certainly used and where debilitating and disabling diseases, may have been much less common. Agriculture seems to have been confined largely to plant cultivation. Very few domesticated animals have a likely rainforest origin; the pig is one notable exception.

It should not be forgotten that vegetation destruction was associated with technological innovation aimed at forest clearance (the introduction, late, of iron tools) drainage of swamps, ponding of upland river valleys to create dyked (bunded) fields, construction of terracing or raised fields both to remove excess water from the land in the hills and mountains of the everwet tropics and to regulate the hydrological regime to provide a more stable moisture supply in the seasonally dry tropics, and construction of artificial field systems to grow crops in lowland locations, as will be discussed later. In many areas the old method of shifting cultivation and some reliance on hunting and foraging continued, as it does today, independently or semi-independently, despite these innovations.

2.2   THE CONSERVATION OF USEFUL PLANTS

If useful plants were selectively conserved before they were deliberately planted or planting and conservation went hand in hand, wild plant food production may have preceded cultivation/agriculture (Spriggs 1993: 141). Obviously. neither the archaeological, geomorphological, pollen or plant macrofossil evidence can adequately sustain this theory. The only support for it is very indirect: microscopic organic residues from taro found on stone tools from the 28,000 year old Kilu Cave site on the island of Buka (north Solomons) suggests that this tuberous plant might have provided a major staple food (Loy *et al.* 1992) and have eventually been taken in to cultivation and from Balof 2 rock shelter, New Ireland, the aroids *Cyrtospermum* or *Alocasia* (Barton and White 1993). Taro could be cultivated without difficult forest clearance as it grows naturally in swamps and along streams.

Most evidence for the conservation of valuable plants by forest peoples is from ethnobotany (cf. Ellen, this volume). However, it is impossible to ignore theoretically considerations, even if they have to be dismissed as speculative or unprovable. Sauer (1952) suggested that agriculture had its origins in the piedmont areas of Southeast Asia and that cultivation of root crops and fruit trees followed on from collecting. There may be some truth in this as the research of Loy *et al.* (1992) indicates, but many more results are needed before most people will be convinced.

Any gardener will tell you that it is easier to grow plants from cuttings, to clone, or from rootstocks, than it is from seed. Sauer contended that this method of reproduction would have been more obvious to early humans than the biological function of seeds. A nice theory, but impossible to prove or disprove because even a consideration of the knowledge and behaviour of existing forest peoples cannot give an answer: argument by analogy is readily refuted. Fruit trees are seed plants so these must, if Sauer's ideas are followed up, have been tended first, then grown from cuttings before the importance of the seeds in reproduction was appreciated.

It is perfectly logical that plants which can reproduce vegetatively should have been used for a very long time but it is informative, turning to Ellen's contribution again, to look at the trees. Some wood can be seen amongst those trees. Perhaps the most intriguing taxon in the Asian context is *Canarium. Canarium* does contribute to the pollen record (Maloney 1996a), but the identification is not completely certain. This writer has found *Canarium* type pollen consistently in highland North Sumatran pollen sites despite the fact that the highest recorded altitude for all but two members of the genus mentioned in *Flora Malesiana* is a few hundred metres lower, but it is never common. There is evidence for possible early forest clearance in the area (see below) but no early archaeological sites, although the archaeology of the region has been exceedingly poorly studied to date. *Canarium* fades from the pollen record at the times of the most intensive forest clearance but reappears, presumably planted, at more recent times.

Macrofossil remains of *Canarium* have been recovered from Hoabinhian and later sites sites in Thailand (Gorman 1969; Yen 1977; Reynolds 1992), remains of wild, but edible fruits, from Sri Lanka , where they date to 12,500-10,000 B.P. (Kajale 1989), Sulawesi (Glover 1985) and Papua-New Guinea (Swadling *et al.* 1990; Yen 1990; Fredricksen *et al.* 1993). Indeed a large fruited variety which Gosden (1993) chose to call domesticated has been found in the Sepik-Ramu area of Papua-New Guinea in an archaeological level dated to 14,000 B.P. and *Canarium* was distributed over a large region by the mid-Holocene while it has been suggested that edible species from the genus may have been introduced to the Bismarck Islands and the Solomons from this source (Wickler 1990).

There is evidence that *Pandanus* has been used for over 12,000 years in Papua-New Guinea (Powell 1982) and may have been selectively conserved. It can be harvested for up to four months in the year (Gorecki 1986), and can be preserved and stored for long periods. A plant of swamp margins, that and the availability of game may have drawn people to the highlands and contributed to the early development of agriculture under cooler conditions than present during the Late Quaternary. While the bottle gourd, *Lagenaria,* a plant which does not contribute to the pollen record, may also have a long history, as it does in the Americas, but there is less evidence for this, although it has been reported (Reynolds 1992) from later prehistoric contexts in Thailand. There is tentative evidence that gambier (*Uncaria*) and pepper (*Piper*) may have been conserved during times of forest destruction (Maloney 1996b) but much more detailed pollen analyses are necessary to confirm this. *Arenga,* one of the sugar palms, might have been selectively conserved or planted in central Sumatra (Morley 1982) but the pollen of this is never common and the Danau Padang pollen record is not well enough dated to confirm if this began 4000 years ago, as has been claimed, or much more recently. Other examples of taxa which may have been selectively conserved could be given.

2.3   THE DEVELOPMENT OF AGRICULTURE: SHIFTING CULTIVATION

The development of agriculture is always dependent upon technological development whether it be the introduction of techniques for removing vegetation without felling, e.g learning how to use fire to manipulate vegetation to the best effect for human purposes, or the use of better implements. The digging stick is the tool most associated with early tropical agriculture and it may have been superceded by the hoe or spades and ultimately, in Asia, the use of the plough, but the tracing of evolutionary sequences (Isaac 1970) is fraught with difficulty and no longer popular with ethnographers, and archaeological finds of wooden artifacts are rare.

How shifting cultivation (swiddening) came to be adopted is shrouded in the mists of time. Swiddening, or slash-burn agriculture, involves clearance of small areas of forest, cultivation of the land for a short period until the soil nutrients are exhausted, then land abandonment leaving the vegetation to regrow. The length of the fallow cycle depends upon population pressure on resources, soil fertility, cultivation practices and the kinds of plants grown. Shifting cultivation may involve periodic abandonment of settlements or permanent settlement and rotational use of the forest. The latter may have evolved from the former, at least in some areas, but this again is difficult to substantiate except by historical analogy. Unfortunately the Kuk site, see below, which has such important evidence for the invention and development of highlands agriculture, has little evidence of settlement (Gorecki 1986).

It has been claimed that shifting cultivation is an ecologically balanced, sustainable use of tropical rainforest land when properly managed and that efforts should be made at understanding and improving it rather than abolition (Cramb 1989) but many writers denounce it as always being destructive rather than constructive, e.g. McGrath (1987). As Hecht *et al.* (1988) indicate, management should not neglect the importance of products from successional fallow to agricultural stability.

It is possible that early peoples recognised that rapid regrowth occurs wherever natural destruction of a rainforest happens and went on from there. However, stone tools would be required to even kill smaller trees by ring barking, but the larger trees may have been left where they were because they were less easy to destroy.

Modern analogies may not be appropriate, but often the smaller trees are felled and then the larger ones are killed by burning, restoring some of the nutrients locked up in the vegetation to the soil.

Most older tropical archaeological sites are limestone caves, like the early Brazilian site (Bahn 1993), or coastal middens, but Late Quaternary open camp sites have been reported from Papua-New Guinea (Gorecki 1986). Coastal shell middens do not normally take a long time to accumulate, and they have been attributed to hunter-gatherers rather than agriculturalists, although middens do sometimes occur where there has been crop cultivation as well as collecting, Khok Phanom Di in central Thailand (Maloney *et al.* 1989) is a case in point. In contrast, cave sites often have long records of continuous or discontinuous occupation and their inhabitants may sometimes have been cultivators, the later occupants of Spirit Cave and Banyan Valley Cave in north west Thailand are an example (Reynolds 1992). However, most shifting cultivator sites may have been as ephemeral as the camps of hunter-gatherers, inseparable from them on the basis of any artifactual remains, and will certainly have been constructed of organic materials which will not preserve in a humid tropical environment.

Palynologists have frequently commented on pollen evidence for forest clearance, but few have examined the microfossil charcoal record to substantiate the possibility of forest destruction by fire, and none have reviewed the methods of swiddening used in historic times to seek clues about what might have happened in the past. Of course, landslides resulting from earthquakes can cause major forest disturbances (Garwood *et al.* 1979) as can hurricanes and windstorms.

The longest continuous fire record from any area of tropical rainforest is from North Queensland (Kershaw 1986) where it extends back to an estimated 179,000 years, suddenly intensifying around 38,000 years, when people were known to have been present in Australia, as Webb (1995) points out, the problem of how to distinguish anthropogenic burning from natural forest fires needs to be addressed.

Natural forest fires dating from about 17,500 to 350 B.P. have been reported from eastern Kalimantan (Goldammer and Seibert 1989) and associated with the intense El Nino - Southern Oscillation (ENSO) event of 1982-83 (Malingreau *et al.* 1985). Mid- to late-Holocene fires have also been reported from areas without known human settlement in Amazonia (Sanford *et al.* 1985).

Hope and Tulip (1994) discovered that the microfossil charcoal curve began after 10,900 B.P. at Lake Hordorli (780 m) in Irian Jaya, and ascribed this to human activity but the first forest clearance is said to be evidenced by the rise of *Casuarina* after 8000 B.P. although (p.394) earlier lowland clearance and regrowth phases are not ruled out. However, it appears from the published pollen diagram that only about 20 pollen samples cover the last 11,000 years of sediment deposition, insufficient for detailed reconstruction of the history of shifting cultivation.

The earliest forest clearance by highland people in Africa seems to have occurred in East Africa, in the Rukiga Highlands of Kenya (Hamilton *et al.* 1986;

Taylor 1990,1993) possibly around 5000 years ago. As is often the case it is not clear what, if anything, was cultivated. Pollen of possible root crop plants is very rarely found anywhere in the world, and macrofossil remains of such plants are not commonly reported from archaeological sites as there is little which can preserve, although (Hather and Kirch 1991; Hather 1994) they do occur occasionally, while most of the grasses from which the cereals come, have pollen grains which cannot be distinguished to the genus let alone the species (Haberle 1994; Maloney 1990, 1994). There are occasional exceptions and maize (*Zea mays*) is often claimed to be one but its phytoliths are more readily identifiable (Bush *et al.* 1992; Pearsall 1994) than the pollen. Nearly all the finds of crop plants from Africa date to after the first millennium A.D. (Shaw 1976) so it is no surprise to find that forest clearance was late compared with the Americas and Southeast Asia. Indeed Maley)1992) suggested that forest clearance only began about 2500-2000 years ago in west Africa but there is dispute even about that as Giresse *et al.* (1994) have interpreted the changes which occurred in palaeoclimatic terms.

Archaeological investigations at various sites on the Santa Maria watershed of Panama in Central America have shown that it has been occupied uninterruptedly for about the last 11,000 years (Cooke and Ranere 1992; Piperno *et al.* 1985, Piperno *et al.* 1991). Here by 8600 B.P. burning of the vegetation was taking place and small clearings in the forest were being made almost certainly for shifting cultivation. Obviously where seasonal dryness occurs burning is likely to play a major role in forest destruction. It is thought that some tuber plants native to the Americas and not generally cultivated today were taken into domestication before 7000 B.P. in Panama. Arrowroot (*Marantia arundinacia*) phytoliths were found in archaeological contexts. However, the most dramatic change occurred with the adoption of a seed plant, maize, around 7000 years ago. Even the introduction of maize did not lead to much forest destruction in the early stages. It led to forest degradation on the hillslopes and the rise of a larger number of small settlements.

There is now much information from plant macrofossil remains, phytoliths and pollen on the origins of maize cultivation in various parts of the Americas from Mexico in the north down to Ecuador in the south (McClung de Tapia 1992; Pearsall 1992; Pearsall and Piperno 1990; Piperno *et al.* 1985; Bush *et al.* 1989; 1992) but there is not so much information on the farming processes involved. It is notable that the Amazon has a 6000 year record of maize cultivation (Bush *et al.* 1989) but that little damage appears to have been done to the forest.

Almost all the highland Southeast Asian pollen diagrams have evidence for disturbance of human or natural origin. The two oldest Late Quaternary pollen records from Indonesia, those from my site of Pea Bullok, (Maloney and McCormac 1995) north of the equator, in North Sumatra, and Danau di-Atas, (Newsome and Flenley 1988) south of the equator in Central Sumatra, extend back 30,000 years. They both show evidence for quite early disturbance of the montane rainforest. At Danau di-Atas it began some time after 8200 B.P. but large scale, regional deforestation only took place after about 1800 B.P. and only five samples covering this time period were analysed. The record is clear cut after 4000 B.P. at Danau Padang (Morley 1982) in the same general region but there is no evidence for what was cultivated although sugar palm (*Arenga*) pollen occurs in the later part of the sequence but it is never abundant, and, in the absence of sufficient radiocarbon dates from the core, may prove to be largely of very recent origin, while archaeological information is scanty.

Pea Bullok is the best dated site from Sumatra with 18 dates in total from two cores, one from the edge of the site, the other from the centre. The base of the core from the edge is around 30,000 years old and that from the centre about 20,000 years old. In both cases peat accumulation has been very slow over the last 12,000 years, 1.5 m. of peat growth at the centre and about 2 m. at the edge. There is evidence for erosion and redeposition of material in the form of a radiocarbon date which is 4000 years too old at 2.2-2.3 m. depth in the core from the edge but this is almost certainly the result of natural causes. A hiccup in forest development also occurs around 9000 B.P. and this is present in a record from a lake site located 15 km. to the east too. In this instance earthquake activity was probably the cause but the data from Kuk suggest that human activity cannot be ruled out.

The core from the centre of Pea Bullok has a record of forest disturbance extending back at least 6000 years ago, while that from the edge indicated that disruption may have begun 8000 years ago (Maloney and McCormac 1995). Agricultural weeds appeared in the pollen sequence for the edge around 6000 B.P. These are more likely to have been associated with the cultivation of dry rice or root crops than wet rice.

The 18,500 year old site of Pea Sim-sim (Maloney 1985, 1996b, Plate 1) shows changes which bear some comparison with the Kuk record in that a swamp forest was cleared 6500 years ago suggesting that dry land forest resources were so exhausted that wood from the swamps was needed. However, analysis of additional pollen samples has revealed that this was not as abrupt as was initially thought. Deforestation was more gradual until *c.* 2500-2000 B.P. The Tao Sipinggan record is more detailed for the last few thousand years and suggests that the final clearance phase, presumably associated with the introduction of wet rice agriculture began about 1700 years ago. All of these sites are at around 1400 m. altitude.

PLATE 1. The swamp at Pea Sim-sim (1400m), North Sumatra, Indonesia.

Pea Sijajap (Maloney 1996b) is at *c*.100 m. lower altitude in the Simamora rift valley. A small patch of swamp was present in an extinct volcanic crater surrounded by rice fields. Only a short core about a metre long was obtained from the site but this proved to be almost 4000 years old. The pollen record was unpromising being dominated by *Eugenia,* a tree, but burnt microfossil remains of grasses were present suggesting that there had been environmental destruction. So the highlands of north and central Sumatra also have a long history of forest disturbance but whether or not there was agricultural development is unclear.

Several Late Quaternary and Holocene pollen diagrams have been published from highland west Java (Stuijts 1993) but separation of natural from human induced vegetation changes is not easy (Maloney 1996b).

Turning lastly to the mainland. It is known from historical records that rice is a late introduction to peninsular West Malaysia (Hill 1977) and the pollen data (Morley 1981) suggests that forest clearance only began about 600 years ago in the east centre of the area.

Haberle *et al.* (1991) suggest that forest clearance predated 7000 B.P. in the upper Baliem Valley of Irian Jaya and that this lead to increased slope runoff, and erosion which explained a break in sediment accumulation at Kelela Swamp between 7000-5200 B.P. Forest clearance was evident again thereafter. At Haeapugua (1650m) in the Tari Basin of the southern Papua-New Guinea highlands (Haberle 1993) the first phase of clearance began about 8000 B.P. and ended around 2000 B.P., then grassland was established and carbonised particle frequencies were high.

While most of the pollen evidence is from the highlands, there is some information from the lowlands. The Markham Valley of Papua-New Guinea is an area of lowland grassland but it is clear from the Lake Wanum (35 m. a.s.l.) pollen record and microfossil charcoal in the sediments that this only became established around 5300 years ago (Garrett-Jones 1979) although disturbance indicators began to increase about 8500 years ago. So the lowlands as well as the highlands were affected by the activities of people but we know less from archaeology about the lowland.

The lowland pollen sites which have been investigated in Indonesia, with the exception of Lake Hordorli in Irian Jaya (Hope and Tulip 1994) are all in freshwater swamp and mangrove swamp areas and show evidence for very localised vegetation changes of natural origin (e.g. Caratini and Tissot 1985, 1988; Muller 1963, 1972; Anderson and Muller 1975). Lake Hordorli is situated at 780m altitude (Hope and Tulip 1994) and disturbance taxa began to increase after 10,500 B.P. and "may include a component from regrowth in clearings in the lowland forests" (p. 394). The carbonised particle curve was matched by a distinct rise in *Casuarina* after 8000 B.P., a taxon often planted in Papua-New Guinea. Haberle *et al.* (1991) suggest that *Casuarina* was widely grown in the upper Baliem Valley of Irian Jaya after 2900 and, especially, 1100 B.P. So, prehistoric people seem to have both deforested and reafforested.

The most detailed evidence for environmental change associated with the activities of people from mainland Southeast Asia is from the lowlands, from the vicinity of Khok Phanom Di in east central Thailand (Maloney *et al.* 1989; Maloney 1991; Kealhofer and Piperno 1994). The archaeological site was occupied from about

4000-2000 B.P. and rice was cultivated throughout. Sedentary agriculture in an estuarine area is indicated. However, rice phytoliths were present over 6000 years ago, but these may have been from a wild species (Kealhofer and Piperno 1994) and the peaks of microfossil charcoal suggest that burning of the mangrove swamp took place in prehistory. In this case it is not certain if this was to clear the land for agriculture or for some other reason but obviously the environment has been disrupted or managed for thousands of years. Kealhofer and myself suspect that some form of shifting cultivation preceded settled agriculture.

For a pollen analyst to be able to distinguish early, ephemeral forest disturbances in the fossil record and reliably assign them to the impact of people is very unlikely. Apart from anything else, this would require close spacing of both pollen and radiocarbon samples as such clearances are likely to have been of short duration. The best way to approach such a task to assess what can be recognised might be to examine short cores of fairly recent origin, as Corlett (1984) has in Papua-New Guinea, in detail. It would also be most appropriate firstly to examine sites from marginal environments, areas where both soils and climatic factors hinder regeneration, as the records of change from these should be more clear cut. It is not surprising that most of the claims for early forest clearances have been made by researchers, including myself, who have worked on sites located in agriculturally marginal areas in the highlands.

2.4.   'DRY LAND' CULTIVATION: BEYOND SIMPLE SWIDDENING

The development of dry land cultivation beyond the shifting cultivation regime involves methods of improving sustainability either in the form of the introduction of new higher-yielding crops plants (e.g. the sweet potato in Papua-New Guinea), improving the varieties of older crop plants or initiation of better crop production methods. At Khok Phanom Di the crop was rice, as indicated above. Dry land cutivation must be distinguished from dry farming. What is meant here is cultivation of crops on dry land in rainforest areas, not dry agriculture.

Foremost among the improved farming techniques has been the introduction of tillage systems to the tropics. Sanoja (1989) claimed that tillage of manioc using the hoe began around 5000 B.P. in the lower Magdalena River Valley of Colombia. The antiquity of tillage in Southeast Asia is unknown but Higham (1989) suggested that the use of the plough was introduced to Bac Bo (Vietnam) by the Han Chinese and that it was they who created the practise of making bunded fields. So, at Khok Phanom Di rice was cultivated without the use of tillage. This, plus poor soils (acid sulphate soils) could help explain the relatively short occupation of the site.

Circumstantial evidence for the introduction of the plough in Asia can be derived from examination of bones from the traction animal, the water buffalo, associated with tillage (Higham *et al.* 1981), but the water buffalo has never been truely domesticated, and a consistent record over time does not exist for Southeast Asia. Groves (1995: 155) states that the oldest putative domestic buffaloe remains come from sites in southern China, but he does not give details. A bronze ploughshare has been reported from Shizhaishan, Yunnan (Higham 1996: 145), but from quite late contexts.

In the Santa Maria watershed agricultural intensification was represented by fewer, larger, settlement sites and the significant changes which occurred in the palaeoecological record around 2500 years ago suggest that sustainability had been achieved and that the communities had become sedentary. This indicates that shifting cultivation no longer involved moving the home base and that some system of rotational land use had evolved. Such detailed records are not available from Southeast Asia yet but the data to hand reveal a surprisingly long length of disturbance by people, if not of settled agriculture. A possible 2500 year plus record for dry land rice cultivation (*ladang*) has been described in Maloney (1996b), however, and archaeologists are beginning to become interested in tracing dry field cultivation (cf. Mudar 1995).

Kuk is in the upper Wahgi valley of the remote Papua-New Guinea highlands at 1580 m above sea level. There were phases at Kuk when swamp agriculture was partly replaced by dryland agriculture (Gorecki 1986) but no implements survive from that time and those of later date comprise of digging sticks and paddle-shaped spades which could not be used for hoeing. Although (Bayliss-Smith and Golson 1992) claim that heavy digging sticks were being used for soil tillage in 1980. The site is not well enough dated to establish if erosion rates have changed significantly over time but Hughes *et al* (1991) suggest that the erosion rate in the catchment has been low by world standards although there was a significant increase in sedimentation rates between 9000-6000 B.P.

Drainage and cultivation of drained land was not confined to the highland swamps of Papua New Guinea in prehistory, but it was much later in the lowlands of Central America where remote sensing of the Maya lowlands has (Adams 1980; Nichols1988; Pope and Dahlin 1989) revealed that swamps were drained, modified and intensively cultivated in a large number of areas. This brings us to the topic of agricultural development in rainforest areas but it should be mentioned that the Maya (Gomez-Pompa 1987) that they also appear to have protected and probably managed their forests as sources of many plant and animal products: they seem to have introduced useful trees, protected wild useful trees, planted trees along field borders, and may have even had cacao plantations. As has happened at least since Hindu-Buddhist times in Asia, sacred trees were probaly conserved. This pattern of land use continues to the present.

Where forest clearance occurs on slopes in humid tropical areas soil erosion is a probability rather than a possibility as Thomas and Nortcliff (this volume) have indicated, and the material removed may accumulate in mid-slope locations as colluvium or be deposited in rivers, lakes, peat bogs or the sea. Soil eroded from the uplands could have been used to construct raised fields in parts of Belize (Jacobs 1995). It is clear from the presence of inwashed minerogenic material in peats that soil erosion has occurred in prehistoric times related to human activities. Kuk has inwashed material attributed to soil erosion which pre-dates 9000 B.P. but Haberle *et al.* (1991) suggested that soil erosion may have begun as early as 28,000 B.P. in the upper Baliem Valley of Irian Jaya.

2.5. DEVELOPMENT OF THE RAINFOREST: HYDROLOGICAL MANIPULATION

While small scale hydrological manipulation like ponding or diversion of small streams may not require excessive input of labour, drainage of swamps and construction of agricultural terracing demands both, and rules about how the water is used. It invariably implies that social power is in the hands of a chief or a governing elite, otherwise such changes would not be possible, and that the population was sedentary. Examination of the origin of swamp drainage and agricultural terracing indicates technological, social and political development as well as the rate of progress made toward agricultural improvement and sustainability in rainforest areas. It may also provide indications of population size at various times in the past (cf. Bayliss-Smith and Golson 1992). Pondfield irrigation is common across Polynesia, as is the type of intensive raised bed cultivation with reticulate drainage/irrigation ditches in swamps found at Kuk, but Kirch and Lepofsky (1993) doubt a Southeast Asian origin and what few radiocarbon dates are available suggest that raised field systems are late.

2. 5. 1. Drainage of swamps
Some crops, e.g. rice, can be grown on swamps without drainage (Fig.1) although minerogenic material from dry land may often be added. Drainage of the swamps at Kuk was thought to have begun around 9000 years ago (Golson 1977; Golson and Hughes 1980; Golson and Gardner 1990; Bayliss-Smith and Golson 1992) and the reliance on the swamps has been interpreted as a response to stress in dry land cultivation, possibly as a result of soil erosion and consequent degradation of the nearby slopes. Unfortunately, pollen records from the region do not extend back this far, so there is no supporting palynological information, or indication of what pollen spectra associated with this kind of presumed disturbance might look like and there are no plant macrofossils to indicate what was cultivated. However, the record of the swamps is very important. A reconsideration of the Mayan subsistence economy (Adams et al. 1981) also concludes that swamps became regarded as assets rather than waste land but (Bloom et al. 1983) they may have been used only as a last resort.

While there is no direct evidence for dryland cultivation at Kuk, the usage of the swamps may simply be a reflection of the kind of plants that they were trying to cultivate. Taro is a possibility but this cannot be proven. There are some indications from the phytolith record (Wilson 1985) that bananas may have been grown but there is no support from the pollen record. Oddly enough, here too agricultural intensification seems to have taken place about 2500-2000 years ago, an obvious response to increased population. This was the time period of the final replacement of forest by grassland in the area. Thereafter swamp management began to take on the form of gridded gardens, suggesting the intensive cultivation of a staple plant, and increasingly advanced agricultural technology. While the start of this cannot be established precisely, its end is marked by a layer of volcanic ash which is c. 1100 years old. What is so important about Kuk as far as we are concerned is not what was grown but that there is evidence for manipulation of wetlands as well as disruption of dryland vegetation in such a remote place so long ago.

2. 5. 2. Terracing
The origins of the many different types of terracing has not been sufficiently investigated on a worldwide scale and terrace structures only rarely yield organic material which can be radiometrically dated. It has been suggested that agricultural terracing began in the Middle East with dry field terraces, and that the idea was diffused westward, southward and eastward (Spencer 1962) merging with the technology of east Asian hydraulic engineering in northern Vietnam and southern China (Spencer 1964) around 2000 B.C. to produce the wet field terrace which, in turn, was also diffused in various directions. It is not possible to examine the various types of terracing in detail

here but Higham (1989, 1996) did not mention terracing from mainland Southeast Asia at all and most information relates to Aztec terracing (Smith and Price 1994), Mexico, Mayan terracing, Belize (Adams 1980; Turner 1974) and Incan terracing in Peru (Hastorf and Earle 1985). Terracing is associated with annual or near-annual cropping, with herbs or shrubs rather than trees.

According to Wheatley (1965), the earliest historical information from Southeast Asia, from Vietnam, dates to the early centuries B.C. and relates to the lowlands of Tonkin, which are unlikely to have had extensive agricultural terracing. In any event the Lac chieftains of that time probably would not have had sufficient power to institute intricate irrigation systems. Reclamation of the deltaic land of that area would have required quite sophisticated hydrological technology and a some form of centralised control. This might have come about when the Chinese invaded Annam in 111 B.C. and bunded fields, a form of incipient terracing, with slight differences in the altitudes of fields, may have come into being some time thereafter, but the written accounts are imprecise. So terracing may be late in the Asian tropics. Irrigation, but not terracing is mentioned in historical accounts from Guatemala and Honduras (Turner 1974). Lowland irrigation and raised field cultivation in Central and South America will be examined in more detail later.

Continuing with Southeast Asia, information about agricultural practises in the upland valleys of Vietnam during the early centuries B.C. is imprecise. There are irrigation channels and stone-embanked terraces of possible ancient origin in the Gio-Linh highlands, which are still used by subsistence farmers, but earlier structures exist for which there seems no known use; narrower, higher terraces, have been abandoned. The situation is no clearer in island Southeast Asia but possibly ancient agricultural terracing is present in places, e.g. the Ifuago area of Luzon in the Philippines. The most detailed recent studies of terracing from this general area, however, seem to be those of the upper Ramu River Valley in the Papua-New Guinea highlands (Sullivan and Hughes 1987) where cut garden terraces on slopes appear to be more than 500 years old.

It is not clear if the innovation of terracing results primarily to lead water off the land, conserve it, counter soil erosion, or solely to intensify crop production but, doubtless, climatic factors played a different role related to both the area and time of introduction.

2. 5. 3. Lowland irrigation features

Drainage is a considerable problem in most tropical everwet lowland areas where the underlying soils are clay-rich. At the more seasonal margins, e.g. in Cambodia, maintenance of a continued water retention for crop, animal and human consumption becomes a problem. It is not surprising that some of the largest feats of prehistoric and protohistoric hydraulic engineering area associated with such areas, in the southern Maya lowlands, as previously mentioned, and around the Tonle Sap in Cambodia (Angkor). The Tonle Sap canals follow a regular patterns (Engelhardt 1996), and there are large storage tanks (barays). Earlier drainage and communication channels associated with Oc Eo in Vietnam (Higham 1989) are less regular. Only the largest Mayan swamp channels could be traced from the air and there is dispute about the patterns shown (Adams 1980; Pope and Dahlin 1989). Adams claimed that most were found in swamps but some occurred at the edges of lakes and lagoons and along rivers. Raised fields were present between the canals and cities were located near this cultivated swamp land although none were within present swamps or inundated zones. A pattern of intensive agriculture with managed wetlands and centralized political control emerges as in Cambodia but Pope and Dahlin (1989) dispute strong dependence on wetland agriculture.

2. 5. 4. Raised fields

Raised (ridged, drained) field agriculture occurs throughout the tropics (Denevan 1970). and seems to be a response to a demand for increased agricultural production (Turner 1974). It is a means of reclaiming swamp land, providing seasonal drainage, weed control, improvement of soil fertility and temperature regulation (Denevan and Turner 1974). This kind of cultivation, known as *chinampas* in central America, and used by both the Mayans and the Aztecs (Turner 1974; Turner and Harrison 1981; Smith and Price 1994) in the Central American lowlands, also occurred in the arid coastal desert of Peru and in high altitude Bolivia around Lake Titicaca (Moore 1988), amongst other parts of the tropical Americas, and its forms and functions vary greatly from region to region. Raised fields of the Bolivian altiplano (Kolata and Ortloff 1989) have been shown to enhance heat storage capacity mitigating against frost damage to maturing crops. The first raised fields to be discovered in the Mayan area were in riverine locations (Turner 1974) but they were later discovered in basins of interior drainage also and maize and amaranths may have been cultivated here, as (Smith and Price 1994) in the Aztec areas and Peru (Moore 1988), but some cacao fragments were recovered from excavations at Kokeal. Macrofossil maize has been recovered from archaeological contexts (Hather and Hammond 1994) in Belize, but roots and tubers were also grown. As at Kuk, the spread of agriculture may have been from dry to wet land but may not have begun before about 2200 B.P. In Peru the earliest raised fields may date to around 900 A.D. (Moore 1988), later than Incan terrace agriculture.

3.   Conclusion

The only region of tropical vegetation that we do not have lengthy evidence of human disturbance for is lowland tropical rainforest but this is probably as much due to the absence of suitable sites for palaeoecological research as anything else. Seasonal forest, montane forest and mangrove swamps have been transformed during prehistory by the activities of people. The clearest effects are often to be found in pollen records from mountain environments, where soil and climatic conditions are often marginal for forest growth anyway, and frequently fire has been an important means of forest destruction. As research proceeds it is likely that the history of human impact on the tropical forest environment and its regional extent will be found to be considerably greater than we suspect at present. It is also likely that more evidence will be found to confirm that the importance of environmental factors has been long recognised and that land use has taken this into account, that there was a skillful use of the land related to its perceived productive capability. The data from the Santa Maria watershed and from Kuk suggest that this was the case.

It is unlikely that we will succeed in detecting the earliest ephemeral forest clearances for agriculture using pollen analysis as they may be inseparable from minor disturbances of natural origin, and apart the exceptions just mentioned, clear evidence for development of agricultural technology in the form that the way land was used over time may prove elusive over wide areas.

References

Adams, R.E.W. (1980). Swamps, canals, and the locations of ancient Maya cities, *Antiquity*, **54**, 206-215.

Adams, R.E.W., Brown, W.E. Jr. and Culbert, T.P. (1981). Radar mapping, archeology, and ancient Maya land use, *Science*, **213**, 1457-1463.

Anderson, J.A.R. and Muller, J. (1975). Palynological study of a Holocene peat and a Miocene coal deposit from N.W. Borneo, *Review of Palaeobotany and Palynology*, **19**, 291-351.

Bahn, P. (1993). 50,000 year old Americans at Pedra Furada, *Nature*, **362**, 114-115.

Bailey, R.C., Head, G., Jenike, M., Owen, B., Rechtman, R. and Zechenter, E. (1989). Hunting and gathering in the tropical rain forest: is it possible?, *American Anthropologist*, **91**, 59-82.

Barton, H. and White, J.P. (1993). Use of stone and shell artifacts at Belof 2, New Ireland, Papua New Guinea, *Asian Perspectives*, **32** (2), 169-181.

Bayliss-Smith, T. and Golson, J. (1992). A Colocasian revolution in the New Guinea highlands ? Insights from Phase 4 at Kuk, *Archaeology in Oceania*, **27**, 1-21.

Bellwood, P. (1990). From late Pleistocene to early Holocene in Sundaland, in C. Gamble and O. Stoffer (eds.), *The World18,000 B.P.* vol. 2, Unwin Hyman, London, pp. 255-263.

Bush, M.B., Piperno, D.R. and Colinvaux, P.A. (1989). A 6000 year history of Amazonian maize cultivation, *Nature*, **340**, 303-305.

Bush, M.B., Piperno, D.R., Colinvaux, P.A., de Oliveira, P.E., Krissek, L.A., Miller, M.C. and Rowe, W.E. (1992). A 14300-yr paleoecological profile of a lowland tropical lake in Panama, *Ecological Monographs*, **62**, 251-275.

Bloom, D.R., Pohl. M., Buttleman, C., Wiseman, F., Covich, A., Miksicek, C., Ball, J. AND Stein, J. (1983). Prehistoric Maya wetland agriculture and the alluvial soils near San Antonio Rio Hondo, Belize, *Nature*, **301**, 417-419.

Caratini, C. and Tissot, C. (1985). Le Sondage Misedor, etude palynologique. *Etudes de Geographie Tropicale,* Centre Nationale de Recherche Scientifique, Bordeaux.

Caratini, C. and Tissot, C. (1988). Paleogeographical evolution of the Mahakam Delta in Kalimantan, Indonesia during the Quaternary and Late Pliocene, *Review of Palaeobotany and Palynology*, **55**, 217-228.

Cherfas, J. (1986). How to grow a rainforest, *New Scientist*, 23 October, 26-27.

Conway, G.R. and McCauley, D.S. (1983). Intensifying tropical agriculture: the Indonesian experience, *Nature*, **302**, 288-289.

Cooke, R. and Ranere, A.J. (1992). Prehistoric adaptations to the seasonally dry forests of Panama, *World Archaeology*, **24**, 114-133.

Corlett, R.T. (1984). Human impact on the subalpine vegetation of Mt. Wilhelm, Papua New Guinea, *Journal of Ecology*, **72**, 841-854.

Cramb, R.A. (1989). Shifting cultivation and resource degradation in Sarawak: perceptions and policies, *Borneo Research Bulletin*, **21** (1), 22-48.

Denevan, W. (1970). Aboriginal drained field cultivation in the Americas, *Science*, **169**, 647-654.

Denevan, W. and Turner, B. II (1974). Forms, functions and associations of raised fields in the Old World Tropics, *Singapore Journal of Tropical Geography*, **39**, 24-33.

Eder, J.F. (1988). Batak foraging camps today: a window to the history of a hunter-gathering economy, *Human Ecology*, **16**, 35-55

Eggert, M.K.H. (1992). The Central African rain forest: historical speculation and archaeological facts, *World Archaeology*, **24**, 1-24.

Endicott, K. and Bellwood, P. (1991). The possibility of independent foraging in the rain forest of Peninsular

Malaysia, *Human Ecology,* **19**, 151-185.

Flenley, J.R. (1994). Pollen in Polynesia: the use of palynology to detect human activity in the Pacific islands, in J. G. Hather (ed.), *Tropical Archaeobotany: Applications and new Developments,* Routledge, London, pp. 202-214.

Fredericksen, C., Spriggs, M. and Ambrose, W. (1993). Pamwak rockshelter: a Pleistocene site on Manus Island, Papua New Guinea, in M.A. Smith, M. Spriggs and B. Fankhauser (eds), *Sahul in Review: Pleistocene Archaeology in Australia, New Guinea and Island Melanesia,* Occasional Papers in Prehistory No. 24, Department of Prehistory, Research School of Pacific Studies, The Australian National University, Canberra, pp.144-152.

Garrett-Jones, S.E. (1979). Evidence for changes in Holocene vegetation and lake sedimentation in the Markham Valley, Papua New Guinea, Unpublished PhD thesis, Australian National University, Canberra.

Garwood, N.C., Janos, D.P. and Brokaw, N. (1979). Earthquake-caused landslides: a major disturbance to tropical forests, *Science,* **205**, 997-999.

Giresse, P., Maley, J. and Brenac, P. (1994). Late Quaternary palaeoenvironments in the Lake Barombi Mbo (West Cameroon) deduced from pollen and carbon isotopes of organic matter, *Palaeoegeography, Palaeoeclimatology, Palaeoecology,* **107**, 65-78.

Glover, I.C. (1985). Some problems relating to the domestication of rice in Asia, in V.N. Misra and P. Bellwood (eds) *Recent advances in Indo- Pacific prehistory,* Oxford and IBH Publishing Co., New Delhi, pp. 265-279.

Goldammer, J.G. and Siebert, B. (1989). Natural rainforest fires in Eastern Borneo during the Pleistocene and Holocene, *Natuurwissenschaften,* **76**, 518-520.

Golson, J. (1977). No room at the top: agricultural intensification in the New Guinea Highlands, in J. Allen, J. Golson and R. Jones (eds), *Sunda and Sahul: Prehistoric Studies in Southeast Asia, Melanesia and Australia,* London, Academic Press, pp. 601-638.

Golson, J. and Gardner, D.S. (1990). Agriculture and sociopolitical organisation in New Guinea Highlands prehistory, *Annual Review of Anthropology,* **19**, 395-417.

Golson, J. and Hughes, P.J. (1980). The appearance of plant and animal domestication in New Guinea, *Journal de la Societe des Oceanistes,* **36**, 294-303.

Gomez-Pompa, A. (1987). On Maya silviculture, *Mexican Studies,* **3**, 1-17.

Gorecki, P.P. (1986). Human occupation and agricultural development in the Papua New Guinea Highlands, *Mountain Research and Development,* **6**, 159-166.

Gorman, C.F. (1969). Hoabinhian: a pebble-tool complex with easrly plant associations in Southeast Asia, *Science,* **163**, 671-673.

Gosden, C. (1993). Understanding the settlement of Pacific islands in the Pleistocene, in M.A. Smith, M. Spriggs and B. Fankhauser (eds.), *Sahul in Review: Pleistocene Archaeology in Australia, New Guinea and Island Melanesia,* Occasional Papers No. 24, Department of Prehistory, Research School of Pacific Studies, Australian National University, Canberra, pp. 131-144.

Groube, L., Chappell, J., Muke, J. and Price, D. (1986). A 40,000 year old human occupation site at Huon Peninsula, Papua New Guinea, *Nature,* **324**, 453-455.

Groves, C.P. (1995) . Domesticated and commensal mammals of Austronesia and their histories, in P. Bellwood, J.J. Fox and D. Tryon (eds.) *The Austronesians: historical and comparative perspectives,* Department of Anthropology, Research School of Pacific and Asian Studies, Australian National University, Canberra, pp. 152-163.

Guidon, N., Pessis, A.-M., Parenti, F., Fontugne, M. and Guerin, C. (1996). Nature and age of the deposits in Pedra Furada, Brazil: reply to Meltzer, Adovasio and Dillehay, *Antiquity,* **70**, 408-421.

Haberle, S. (1993). Pleistocene vegetation change and early human occupation of a tropical mountainous environment, in M.A. Smith, M. Spriggs and B. Fankhauser (eds), *Sahul in Review: Pleistocene Archaeology in Australia, New Guinea and Island Melanesia,* Occasional Papers in Prehistory, No. 24. Department of Prehistory, Research School of Pacific Studies, Australian National University, Canberra, pp. 109-122,

Haberle, S. (1994). Anthropogenic indicators in pollen diagrams: problems and prospects for late Quaternary palynology in New Guinea, in J. Hather (ed), *Tropical Archaeobotany: Applications and New Developments*, Routledge, London, pp. 172-201.

Haberle, S. G., Hope, G.S. and DeFretes, Y. (1991). Environmental change in the Baliem Valley, montane Irian Jaya, Republic of Indonesia, *Journal of Biogeography*, **18**, 25-40.

Hamilton, A.C., Taylor, D. and Vogel, J.C. (1986). Early forest clearance and environmental deterioration in South-West Uganda, *Nature*, **320**, 164-167.

Harrison, P. (1977). Indonesia: food, population, land. Can Indonesia farm the swamps?, *New Scientist*, 22/29 December, 804-805.

Hart, T.B. and Hart, J.A. (1986). The ecological basis of hunter-gatherer subsistence in African rain forest: the Mbuti of Eastern Zaire, *Human Ecology*, **15** (4), 463-491.

Hastorf, C. and Earle, T. (1985). Intensive agriculture and the geography of political change in the Upper Mantaro region of Peru, in I. Farrington (ed.), *Prehistoric Intensive Agriculture in the Tropics. British Archaeological Reports (International Series)* 232. British Archaeological Reports, Oxford, pp. 569-595.

Hather, J. G. (1994). The identification of charred root and tuber crops from archaeological sites in the Pacific, in J. G. Hather (ed), *Tropical Archaeobotany: Applications and New Developments*, Routledge, London, pp. 51-64.

Hather, J.G. and Hammond, N. (1994). Ancient Maya subsistence diversity: root and tuber remains from Cuello, Belize, *Antiquity*, **68**, 330-335.

Hather, J. G. and Kirch, P. V. (1991). Prehistoric sweet potato (*Ipomoea batatas*) from Mangaia Island, central Polynesia, *Antiquity*, **65**, 887-893.

Headland, T.N. (1987). The wild yam question: how well could independent hunter-gatherers live in a tropical rain forest ecosystem? *Human Ecology*, **15** (4), 463-491.

Headland, T.N. and Reid, L.A. (1989). Hunter-gatherers and their neighbours from prehistory to the present, *Current Anthropology*, **30** (1), 43-66.

Hecht, S.B., Anderson, A.B. and May, P. (1988). The subsidy from nature: shifting cultivation, successional palm forests, and rural development, *Human Organisation*, **47**, 25-34.

Higham, C. (1989). *The Archaeology of Mainland Southeast Asia*, Cambridge University Press, Cambridge

Higham, C. (1996). *The Bronze Age of Southeast Asia*, Cambridge University Press, Cambridge.

Higham, C.F.W., Kijngam, A., Manly, B.F. and Moore, S.J. (1981). The bovid third phalanx and prehistoric ploughing, *Journal of Archaeological Science*, **6**, 353-365.

Hill, R.D. (1977). *Rice in Malaya: a Study in Historical Geography*, Oxford University Press, Kuala Lumpur.

Holden, C. (1986). Regrowing a dry tropical rainforest, *Science*, **234**, 809-810.

Hope, G.S. and Tulip, J. (1994). A long vegetation record from lowland Irian Jaya, Indonesia, *Palaeoegeography, Palaeoclimatology, Palaeoecology*, **109**, 385-398.

Hughes, P.J, Sullivan, M.E. and Yok, D. (1991). Human-induced erosion in a highlands catchment in Papua-New Guinea: the prehistoric and contemporary records, *Zeitschrift fur Geomorphologie*, **83**, 227-239.

Hyndman, D. (1988). Ok Tedi: New Guinea disaster mine, *The Ecologist*, **18** (1), 24-29.

Isaac, E. (1970). *Geography of Domestication*, Prentice-Hall, Eaglewood Cliffs, New Jersey.

Jacobs, J.J. (1995). Ancient Maya wetland agricultural fields in Cobweb Swamp, Belize: construction, chronology, and finction, *Journal of Field Archaeology*, **22**, 175-190.

Kajale, M.D. (1989). Mesolithic exploitation of wild plants in Sri Lanka: archaeobotanical study at the cave site of Beli-Lena, in D.R. Harris and G.C. Hillman (eds), *Foraging and Farming: the Evolution of Plant Domestication*, Unwin Hyman, London, pp.269-281.

Kajale, M.D. (1994). Archaeobotanical investigations on a multicultural site at Adam, Maharashtra, with special reference to the development of tropical agriculture in parts of India, in J.G. Hather (ed.), *Tropical Archaeobotany: Applications and new Developments*, Routledge, London, pp. 34-50,.

Kartinawinata, K., Adiosoemartono, S., Riswan, S. and Vadya, A.P. (1981). The impact of man on a tropical forest in Indonesia, *Ambio*, **10** (2-3), 115-119.

Kealhofer, L. (1996). The human environment during the terminal Pleistocene and Holocene in Northeastern Thailand: preliminary phytolith evidence from Lake Kumphawapi, *Asian Perspectives*, **35** (2), 229-254.

Kealhofer, L. and Piperno, D.R. (1994). Early agriculture in Southeast Asia: phytolith evidence from the Bang Pakong Valley, Thailand, *Antiquity*, **68**, 564-572.

Kershaw, A.P. (1986). Climatic change and Aboriginal burning in north-east Australia during the last two glacial/interglacial cycles, *Nature*, **322**, 47-49.

Kirch, P.V. and Lepofsky, D. (1993). Polynesian irrigation: archaeological and linguistic evidence for origins and development, *Asian Perspectives*, **32** (2), 183-204.

Kolata, A.L. and Ortloff, C. (1989). Thermal analysis of Tiwanaku raised field systems in the Lake Titicaca Basin of Bolivia, *Journal of Archaeological Science*, **16**, 233-263.

Loy, T.H., Spriggs, M. and Wickler, S. (1992). Direct evidence for human use of plants 28,000 years ago: starch residues on stone artifacts from the northern Solomon Islands, *Antiquity*, **66**, 898-912.

Maley, J. (1992). Mise en evidence d'une pejoration climatique entre ca. 2500 et 2000 ans B.P. en Afrique tropicale humide, *Bulletin de Societe Geologique France*, **163**, 363-365.

Malingreau, J.P., Stephens, G. and Fellows, L. (1985). Remote sensing of forest fires: Kalimantan and North Borneo in 1982-83, *Ambio*, **14** (6), 314-321.

Maloney, B.K. (1985). Man's impact on the rainforests of West Malesia : the palynological record, *Journal of Biogeography*, **12**, 537-558.

Maloney, B.K. (1990). Grass pollen and the origins of rice agriculture in north Sumatra, *Modern Quaternary Research in Southeast Asia*, **11**, 135-161.

Maloney, B.K. (1991). Palaeoenvironments of Khok Phanom Di: the pollen, pteridophyte spore and microscopic charcoal record, in C.F.W. Higham and R. Bannanurag (eds), *The Excavation of Khok Phanom Di, Vol.2 (Part I) The Biological Remains,* Research reports of the Society of Antiquaries. Society of Antiquaries, London, pp.7-134.

Maloney, B.K. (1993). Palaeoecology and the origin of the coconut, *Geojournal*, **31**, 355-362.

Maloney, B.K. (1994). The prospects and problems of using palynology to trace the origins of tropical agriculture: the case of Southeast Asia, in J. G.Hather (ed), *Tropical Archaeobotany: Applications and New Developments,* Routledge, London, pp. 139-171.

Maloney, B.K. (1996a). *Canarium* in the Southeast Asian and Oceanic archaeobotanical and pollen records, *Antiquity*, **70**, 926-933.

Maloney, B.K. (1996b). New perspectives on possible early dry land and wet land rice cultivation from highland north Sumatra. Centre for South-East Asian Studies, University of Hull, Occasional Paper No. 29.

Maloney, B.K. and McCormac, F.G. (1995). Thirty thousand years of radiocarbon dated vegetation and climatic change in highland Sumatra., *Radiocarbon*, **37**, 181-190.

Maloney, B.K., Higham, C.F.W. and Bannanurag, R. (1989). Early rice cultivation in Southeast Asia : archaeological and palynological evidence from the Bang Pakong Valley, Thailand, *Antiquity*, **63**, 363-370.

Meltzer, D.J., Adovasio, J.M. and Dillehay, T.D. (1994). On a Pleistocene human occupation at Pedra Furada, Brazil, *Antiquity*, **68**, 695-714.

McClung de Tapia, E. (1992). The origins of agriculture in Mesoamerica and Central America.. in C.W. Cowan and P.J.Watson (eds), *The Origins of Agriculture: an International Perspectives*, Smithsonian Institution Press, Washington, pp. 143-171.

McGrath, D.G. (1987). The role of biomass in shifting cultivation, *Human Ecology*, **15** (2), 221-242.

Moore, J.D. (1988). Prehistoric raised field agriculture in the Casma Valley, Peru, *Journal of Field Archaeology*, **15**, 265-276.

Morley, R.J. (1981). The palaeoecology of Tasek Bera, a lowland swamp in Pahang, west Malaysia, *Singapore Journal of Tropical Geography*, **2**, 49-56.

Morley, R.J. (1982). A palaeoecological interpretation of a 10,000 year old pollen diagram from Danau Padang, Central Sumatra, Indonesia, *Journal of Biogeography*, **9**, 151-190.

Mountain, M-J. (1991). Landscape use and environmental management of tropical rainforest by pre-agricultural hunter-gatherers in northern Sahulland, *Bulletin of the Indo-Pacific Prehistory Association*, **11**, 54-68.

Mudar, K. M. (1995). Evidence for prehistoric dryland farming in mainland Southeast Asia: results of regional survey in Lopburi Province, Thailand, *Asian Perspectives*, **34**, 157-194.

Muller, J. (1963). Palynological study of Holocene peat in Sarawak. *Symposium on ecological research in humid tropics vegetation, Kuching, Sarawak, July 1963,* UNESCO, pp. 147-156.

Muller, J. (1972). Palynological evidence for change in geomorphology, climate and vegetation in the Mio-Pliocene of Malesia, in P. and M. Ashton (eds), *The Quaternary era in Malesia, Transactions of the Second Aberdeen-Hull Symposium on Malesian Ecology, Aberdeen, 1971,* University of Hull, Department of Geography, Miscellaneous Series No. 13, Hull, pp. 6-16,.

Nepstad, D.C., de Carvalho, C.R., Davidson, E.A., Jipp, P.H., Lefebvre, P.A., Negreiros, G.H., da Silva, E.D., Stone, T.A., Trmbore, S.E. and Vieira, S. (1994). The role of deep roots in the hydrological and carbon cycles of Amazonian forests and pastures, *Nature*, **372**, 666-669.

Newsome, J. and Flenley, J.R. (1988). Late Quaternary vegetational history of the Central Highlands of Sumatra. II. Palaeopalynology and vegetational history, *Journal of Biogeography*, **15**, 363-386.

Nicholls, D.L. (1988). Infrared aerial photography and prehispanic irrigation at Teotihuacan: the Tlajinga canals *Journal of Field Archaeology*, **15**, 17-27.

Parenti, F.N., Mercer, N. and Valladas, H. (1990). The oldest hearths of Pedra Furada, Brasil: thermonluminescence analysis of heated stones. *Current Research in the Pleistocene*, **7**, 36-38.

Pearsall, D.M. (1992). The origins of plant cultivation in South America, in C.W. Cowan and P.J. Watson (eds), *The Origins of Agriculture: an International Perspective*, Washington, Smithsonian Institution Press, pp. 173-205.

Pearsall, D.M. (1994). Investigating New World tropical agriculture: contributions from phytolith analysis, in J. G.Hather (ed), *Tropical Archaeobotany: Applications and New Developments,* Routledge, London, pp. 115-138.

Pearsall, D.M. and Piperno, D.R. (1990). Antiquity of maize cultivation in Ecuador: summary and reevaluation of the evidence, *American Antiquity*, **55**, 324-337.

Piperno, D.R., Bush, M.B. and Colinvaux, P.A. (1991). Paleoecological perspectives on human adaptation in Central Panama. II. The Holocene, *Geoarchaeology*, **6**, 227-250.

Piperno, D.R., Husum-Clary, K., Cooke, R.G., Ranere, A.J. and Weiland, D. (1985). Preceramic maize in central Panama: evidence from phytoliths and pollen, *American Anthropologist*, **87**, 871-878.

Pope, K.O. and Dahlin, B.H. (1989). Ancient Maya wetland agriculture: new insights from ecological and remote sensing research, *Journal of Field Archaeology*, **16**, 87-106.

Possell, G. and Kennedy, K. (1979). Hunter-gatherer/agriculturalist exchange in prehistory: an Indian example, *Current Anthropology*, **20**, 592-593.

Powell, J.M. (1982). Plant resources and palaeobotanical evidence for plant use in the Papua New Guinea Highlands, *Archaeology in Oceania*, **17** (1), 28-37.

Reynolds, T.E.G. (1992). Excavations at Banyan Valley Cave, Northern Thailand: a report on the 1972 season, *Asian Perspectives*, **31**, 117-132.

Sanford, R.L. Jr., Saldarriaga, J., Clark, K.E., Uhl, C. and Herrera, R. (1985). Amazon rain-forest fires, *Science*, **227**, 53-55.

Sanoja, M. (1989). From foraging to food production in northeastern Venezueal and the Caribbean, in D.R. Harris and G.C. Hillman (eds), *Foraging and Farming: the Evolution of Plant Exploitation*, Unwin Hyman, London, pp. 523-537.

Sauer, C.O. (1952). *Agricultural Origins and Dispersals*, The American Geographical Society, New York.

Secrett, C. (1986). The environmental impact of transmigration, *The Ecologist*, **16** (2/3), 77-88.

Shaw, T. (1976). Early crops in Africa, in J.R. Harlan, J.M.J. de Wet and A. Stemler (eds), *Origins of African Plant Domestication*, Mouton, The Hague, pp. 107-153.

Smith, M.E. and Price, T.J. (1994). Aztec-period agricultural terraces in Morelos, Mexico: evidence for household-level agricultural intensification, *Journal of Field Archaeology*, **21**, 169-179.

Spencer, J.E. (1962). The origin, nature, and distribution of agricultural terracing, *Pacific Viewpoint*, **2** (1), 1-40.

Spencer, J.E. (1964). The development and spread of agricultural terracing in China, in S.G. Davis (ed.), *Symposium on land use and mineral depositis in Hong Kong, Southern China and South East Asia*, University Press, Hong Kong, pp. 105-110.

Stiles, D. (1993). Hunter-gatherer trade in wild forest products in the early centuries A.D. with the port of Broach, India, *Asian Perspectives*, **32**, 153-167.

Stuijts, I-L. M. (1993). Late Pleistocene and Holocene vegetation of west Java, Indonesia, *Modern Quaternary Research in Southeast Asia*, **12**, 1-173.

Sullivan, M. and Hughes, P. (1986). The geomorphic setting of prehistoric garden terraces in the Eastern Highlands of Papua New Guinea, in Gardiner, V. (ed.), *International Geomorphology 1986, Part II*, pp. 569-582.

Suter, K. (1982). *East Timor and West Irian*, Minority Rights Group Report No. 42. Minority Rights Group, London.

Swadling, P., Araho, N. and Ivuyo, B. (1991). Settlements associated with the inland Sepik-Ramu sea, *Bulletin of the Indo-Pacific Prehistory Association*, **11**, 92-116.

Taylor, D.M. (1990). Late Quaternary pollen records from two Ugandan mires: evidence for environmental change in the Rukiga Highlands of southwest Uganda, *Palaeogeography, Palaeoclimatology, Palaeoecology*, **80**, 283-300.

Taylor, D.M. (1993). Environmental change in montane southwest Uganda : a pollen record for the Holocene from Ahakagyezi Swamp. *The Holocene*, **3**, 324-332.

Thompson, G.B. (1994). Wood charcoals from tropical sites: a contribution to methodology and interpretation, in J. G. Hather (ed.), *Tropical Archaeobotany: Applications and new Developments*, Routledge, London, pp. 9-33.

Turner, B.L. II. (1974). Prehistoric intensive agriculture in the Mayan lowlands, *Science*, **185**, 118-124.

Turner, B.L. II and Harrison, P.D. (1981). Prehistoric raised-field agriculture in the Maya lowlands, *Science*, **192**, 399-405.

Webb, R.E. (1995). ODP Site 820 and the initial human colonisation of Sahul, *Quaternary Australasia*, **13** (1), 13-18.

Wheatley, P. (1965). Agricultural terracing, *Pacific Viewpoint*, **6** (2), 123-144.

Whitten, A.J. (1987). Indonesia's transmigration program and its role in the loss of tropical rainforests, *Conservation Biology*, **1** (3), 239-246.

Wickler, S. (1990). Prehistoric Melanesian exchange and interaction: recent evidence from the northern Solomon Islands, *Asian Perspectives*, **29**, 135-154.

Wilson, S. (1985). Phytolith analysis at Kuk, an early agricultural site in Papua New Guinea, *Archaeology in Oceania*, **20**, 90-9

Wilson, W.L. and Johns, A.D. (1982). Diversity and abundance of selected animal species in undisturbed forest, selectively logged forest and plantations in East Kalimantan, Indonesia, *Biological Conservation*, **24**, 205-218.

Yen, D.E. (1977). Hoabinhian horticulture? The evidence and questions from northwest Thailand, in J. Allen, J. Golson and R. Jones (eds), *Sunda and Sahul: prehistoric studies in Southeast Asia, Melanesia and Australia*, Academic Press, London, pp. 567-599.

Yen, D.E. (1990). Environment, agriculture and the colonisation of the Pacific, in D.E. Yen and J.M.J. Mummery (eds), *Pacific Production Systems,* Occasional Papers in Prehistory No.18, Department of Prehistory, Research School of Pacific Studies, The Australian National University, Canberra, pp. 258-277.

Dr. Bernard K. Maloney, Palaeoecology Centre, The Queen's University, Belfast BT7 1NN, Northern Ireland

5.   INDIGENOUS KNOWLEDGE OF THE RAINFOREST
        Perception, extraction and conservation

        Roy F. Ellen

1.   Introduction

Indigenous knowledge is currently flavour of the month: both economic commodity and
political slogan. It has a market value placed upon it, and has become pivotal in
preserving the identity and culture of indigenous peoples whose traditional way of life
is under threat [1]. In this chapter it is intended to review how rainforest peoples
conceptualise their interactions, construct their ethnobiological knowledge and alter and
maintain the character of the forest through their activities. What will be said will
substantiate the observation that indigenous peoples have perceived, interacted with,
and made use of tropical rainforest in historically diverse ways, and that this diversity
has sometimes been obscured by the understandable prominence given to the
experiences of particular peoples with a high political profile, such as the Kayap'o,
Yanomami and Penan [2]. This process of globalising particular instances has resulted in
an over-simplification of the relationships which people can establish with forest. It
will be argued that it is important for those making recommendations in the fields of
forest management to take indigenous knowledge seriously, but also to form balanced
judgements based on evidence available for particular situations.

2.   Domesticating the forest

We are all sometimes persuaded to think that rainforest is a fragment of some vast
unchanging past which has intruded into the present. In the popular imagination,
peoples of the tropical rainforest are remote, isolated, living in more or less the same
place, unchanging, 'in harmony' with their surroundings. In fact, we now have plenty of
evidence to the contrary, and although the rainforest does indeed have a long ecological
history, it is far from stable and unchanging [3]. Moreover, its history, at least for the last
10,000 years, has also been a cultural history: not only the context in which human
social and ecological change has taken place but an environment which humans may
have been instrumental in, by turns, maintaining and altering. Whether through simple
extraction or low intensity farming, the cumulative long term effects of these
disturbances on forest composition and structure, compared with those of other large
mammals living at similar densities must have been considerable, at least in some
areas[4]. This has improved the rainforest as a human resource base and contributed to its
structural patchiness and biodiversity, and persistent interventions over many hundreds
of years have had important co-evolutionary consequences.

        Long-term human impact has taken various forms, and we can obtain some
measure of it by examining ethnographic evidence drawn from what we know about
contemporary and historically recent food collectors and small-scale agriculturalists.
Even groups subsisting at low population densities modify their habitats by increasing,
say, river sediment loads as a result of agricultural soil disturbance and erosion;
introduction of humanly-transmitted pathogens and other toxins, by changing soil
nutrient levels, disturbing structure and causing erosion (Rambo 1985: 58, 63). Humans

87

B.K. Maloney (ed.), Human Activities and the Tropical Rainforest, 87-99.
© 1998 *Kluwer Academic Publishers. Printed in the Netherlands.*

alter the forest inadvertently by helping disseminate certain seeds of wild plants (abandoned camps, gardens, and villages providing particularly good examples of this), while anthropogenic secondary growth may constitute habitats for new kinds of plants and grazing animals. Even small groups of hunter-gatherers may change their habitat by dropping selected seeds which they collect for food. In this way the Mbuti of the Ituri forest in Zaire propogate genera such as *Canarium* and *Landolphia* (Ichikawa 1992; cf. Fox 1953). Clearance for temporary cultivation plots not only transforms forest structure through cultivation itself, and through regrowth, but also through the selective removal of trees. Large trees with hard woods have selective advantage in being more difficult to remove. On Seram, to the east of the Indonesian archipelago, the presence, for example, of *Canarium vulgare, Sterculia* and *Diospyros ebenaster,* pose difficulties for Nuaulu cultivators [5]. However, plants may be preserved deliberately as well as by default, and many techniques are reported which involve degrees of protection of otherwise wild species (Ellen 1994: 205-206; Headland 1987; Rambo 1985: 71). Collection of forest products specifically for trade (particularly resins, rattans and seeds) has probably been a major selection pressure in the Malaysian peninsula (Dunn 1975; Gianno 1990; Rambo 1979: 60). Human settlement has led to the deliberate introduction of plant domesticates from other parts of the world and many varieties of cultivated trees (Fox 1953; Rambo 1985: 70). The magnificent *Tectona grandis* is now well established in the lowland forest of Seram, though it was probably introduced during the seventeenth century (Ellen 1985: 563). In some parts of Southeast Asia quickly growing species are planted in plots to ensure rapid and appropriate regrowth, and to supply fuel (Whitmore 1990: 135; Barrow, this volume).

Thus, we must be clear that when we seek to 'preserve' rainforest, we choose between preserving the rainforest as it has evolved over the past 10,000 years (including its human component) and changing it by keeping humans out.

3.   **Ways of human life in the rainforest**

In terms of impact on rainforest ecology, to distinguish between the effects of those non-agricultural forms of human extraction we call hunter-gathering and low intensity agriculture is sometimes rather difficult [6]. There is now plenty of evidence for the manipulation and regulation of plant resources in otherwise food collecting populations of the rainforests in ways which maintain or increase yield (Hutterer 1983: 173). For example, replanting the heads of wild yams and protecting valuable fruit-bearing trees; the deliberate burning of bamboo clumps in order to facilitate the extraction of desirable haulms and to promote the growth of green shoots (Rambo 1985: 70). Those peoples engaged in 'wild' palm sago extraction, extract selectively, detach and protect suckers thrown out by mature palms and exercise certain forms of ownership (Ellen 1988). Often, such activity is sufficiently organised, purposeful and significant to warrant the description 'rainforest management'. At what point management becomes cultivation is a major scientific puzzle. The Huastec of the Sierra Madre in Mexico use 63% of the 800 wild species recorded and 25% are actively manipulated (Alcorn 1981: 410). The selective extraction of wild species, strategic burning, and swiddening at optimal conditions may combine to give rise to distinctive patches and new opportunities for colonisation: and there are numerous examples - such as that provided by the Kayap'o (Posey 1988: 89) - of deliberate preservation of corridors of mature forest between plots as some kind of biological reserve.

Those peoples anthropologists conventionally label 'hunters and gatherers' often do things other than hunting and gathering; indeed practices which may assume a critical position in terms of identity and ideology may be rather unimportant in terms of objective ecological measurements (e.g. Barnard 1983). Some groups are similar to agriculturalists from an ecological point of view in the ways that they extract, protect and ensure future supplies of plant resources (e.g. Hutterer 1983: 173, 176; Posey 1982; Rambo 1985). Indeed, whether or not the rainforest could ever have supplied the carbohydrate requirements of food collectors without cultivation has been seriously questioned (Headland 1987; Headland and Bailey 1991). According to this view a key adaptive role must have been played by energy subsidies obtained through exchange. Trade and exchange have existed for centuries between interior or up river peoples including remote foraging populations, linking them to the peoples of the forest fringes, the estuaries or coasts, and ultimately the global economy (Dunn 1975; Hoffman 1984). Thus, such populations are not only involved in collecting forest products which enter the world system, but may be dependent upon inputs from non-food collecting groups for their biological and social survival. This has been the case for many hundreds of years for peoples as diverse as the Agta of the Northern Philippines and the Baka of Cameroun [7].

4.   Ethnobiological knowledge

The question now arises as to what all this has to do with ethnobiological knowledge, by which is meant what people untutored in science, know about plants and animals. It is recognised, of course, that individual subsistence techniques, and therefore different overall combinations of strategies employed by particular populations, have different ecological profiles: in terms of energy transfer, limiting factors and carrying capacity, the degree of human effort required, their effects on the landscape, and the cultural regulation of environmental relations. However, by the same token, they must presumably also have different knowledge profiles. The successful adaptation of humans to rainforest environments rests on their ability to maintain population-land ratios at a level which will permit sustainable extraction, which in turn depends on their capacity to organise and apply knowledge of rainforest structure and composition (Ellen: in press).

What do we mean by an 'ethnobiological knowledge profile'; and how might we begin to measure it? How can we access it, and compare it from one group to another? We can begin by considering the main components of this knowledge, common to all rainforest peoples [8]. These are:

1.   species-focussed, empirical knowledge of individual organisms.

2.   knowledge of general principles based on the observation of many different species.

3.   knowledge of systems of interconnected organisms.

1.   Species-focussed, empirical knowledge of individual organisms.

This includes information on the form, physiology, behaviour, feeding habits, connections with other species of individual organisms, activities of predators and the effects of diseases. Such knowledge is highly variable in relation to each organism. The Nuaulu, for example, know an enormous amount about the wild pig, enough to fill a short monograph, whereas few people have more than a passing acquaintance with worm snakes, which they rarely see and have little interest in (Ellen 1993a: compare 36-39 with 103). Those who have attempted to measure degree of utility, have conclusively shown that it is very asymmetrically distributed with respect to named species, and that it is only when we examine our data in this way that we can see that what constitutes a 'use' is highly problematic (Hays 1982).

Some of this kind of knowledge will obviously be adaptive in marginal situations, i.e. where selective pressure is at its maximum. This information is most likely to include understanding of reproductive biology of particular organisms, of which parts are useful, and how to process them, of the damage that they can do to humans and other organisms upon which humans are dependent (e.g. the effect of toxic yams, insect pests, dangerous snakes, and so on), the role of organisms in dispersing seed, and the use made of the species by non-human organisms as food.

The Nuaulu, for instance, are well informed about many species not because they are of direct value to humans but because they represent the food of animals which they hunt, particularly the cassowary, pig and deer. So, plants and animals should not be viewed in isolation but have to be understood as part of the web of forest life. It is adaptively more important to distinguish varieties of yam from one another because one contains toxic levels of dioscorine and the other is edible than to distinguish them on the basis of perceptual criteria alone (say size of leaf), although such features may indicate crucial functional conditions (Boster 1984). This kind of knowledge is the result of generations of accumulated experience, experimentation and information exchange (Boster 1986; Posey *et al.* 1984; Richards 1985).

Ethnobiological studies of rainforest peoples have, however, uniformly demonstrated an impressive breadth, as well as *depth,* of knowledge of particular significant species. Recent attempts to collate data on the total inventories of plant categories for different subsistence populations shows strikingly how rainforest populations have repeatedly been found to yield much longer lists than populations living in other environments, lists consisting of between 800 and 2,000 items (Berlin 1992; Brown 1985: 44; Ellen: in press). This, of course, reflects relative biodiversity, which we know constrains ethnobiological inventories markedly, such that inventories of tropical rainforest agricultural and non-agricultural peoples tend to be more similar than those of of agriculturalists and non-agriculturalists globally. To some extent it also reflects the subsistence necessity of those who extract from such environments. Though, as we have seen above, it is still debated whether or not human populations could ever have entirely survived on rainforest plant matter without cultivation, breadth of knowledge is undoubtedly a key part of any adaptive strategy.

2.   Knowledge of general principles based on observation of many species.

This may be more important in the long run than either breadth of formal knowledge or depth of substantive knowledge of individual organisms. It is quite clear that general ethnobiological lessons are learned from observing particular instances, and that in the transmission of knowledge overarching deductive models of how the natural world works are privileged over accumulated inductive knowledge. Thus, both pre-emptive and retrospective control of resources are well understood by food collectors as well as cultivators (Ellen 1994: 204). The evidence for regulation of rainforest resources by food collectors suggests, along with the pre-adaptation of knowledge and equipment, that the cultural preconditions for the emergence of agriculture were present long before its existence as a major mode of subsistence. The main elements of agriculture, individually or combined, are all known for so-called pre-agricultural systems, with the possible exception of seed selection and artificial dispersal (Yen 1989: 57).

3.   Knowledge of systems of interconnected organisms.

It has become clear that systematic encyclopaedic knowledge is situated within general folk models, which reflect an ability to connect observations at the species level with informed perceptions about forest structure and dynamics. Thus, Posey (1988) has shown how the Kayap'o of Amazonia maintain buffer zones between gardens and forest which contain plants with nectar-producing glands on their foliage which have the effect of drawing away aggressive ants and parasitic wasps from crops. Inevitably, we can expect what is regarded as 'forest' to vary cross-culturally, but even if we restrict ourselves to focal shared meanings it is clearly complex (e.g. Ellen 1993b; Dwyer 1996). So, for the Nuaulu forest is anything but uniform or empty in the way that they perceive, understand and respond to it. It is more like a mosaic of resources, and a dense network of particular places each having different material values. In this sense it is much like the modern scientific modelling of rainforest as a continuous aggregation of different biotopes and patches, varying according to stages in growth cycles and degree of regeneration. Viewed this way, conventional, distinctions between 'secondary' and 'primary' forest begin to look pretty academic. Seventy eight percent of the 272 forest trees identified by the Nuaulu have particular human uses, and it is through their uses that they are apprehended (Ellen 1985b), wherever they are found. The picture is similar elsewhere. On average, the Panare, Tembe, Urobu and Chacobo peoples of the Amazonian basin use at least two thirds of the tree species growing in the forest (Carneiro 1978, 1988: 79). Such peoples, as with the Indonesian Nuaulu, being forest-fallow swidden cultivators, also have sensitive understandings of how forest changes as a consequence of soil differences, selective extraction, cutting and burning; and of the regrowth stages following abandonment. It is this knowledge which permits such strategies - despite persistent rumours - to be self-sustaining (Dove 1983a, b), and which underly their deliberate application in ways which assist the recovery of degraded areas (Conklin 1957; Sillitoe 1991).

5.   The cultural embeddedness of technical knowledge

Such empirical knowledge of plants and animals as has already been referred to does not exist apart from a broader socially informed understanding of the world, in some kind of hermetically sealed vacuum from which other aspects of culture are excluded. Detailed knowledge of plant reproduction or symbiosis may, for example, comfortably co-exist with beliefs about the world which have not been empirically tested in a conventional scientific sense. Everything is seen as connected through claims of mutual causation to give rise to a complex notion of nature. Indeed, it is often plausibly argued that rainforest peoples have cosmologies which in certain respects anticipated the systems view of the world which underlies modern ecology; as has been the case for the Tukano of the northwest Amazon (Reichel-Dolmatoff 1976). For the Nuaulu, no less than for the Tukano, forest is never experienced as homogenous, rather it is a complex category connecting notions of history, identity and place to pragmatic subsistence concerns. It is also highly-charged morally (Ellen 1993b; cf. Richards 1992).

Although uncut forest is recognised by the Nuaulu as a single entity, it contrasts in different ways with other land types depending on context. It may contrast with owned land, which may sometimes display very mature forest growth, emphasising a jural distinction; with garden land, emphasising human physical interference; or with village land, emphasising landforms: empty as opposed to well-timbered space, inhabited as opposed to ininhabited space, untamed as opposed to tamed space, all with various symbolic associations and practical consequences for Nuaulu consumers. There are no Nuaulu words for either 'nature' or 'culture' but it is in these various and aggregated senses that the Nuaulu come closest to having such a term, and from which the existence of an abstract covert notion of 'nature' can reasonably be inferred. The values with which the Nuaulu invest forest are thus multi-faceted and, in the same way that the material uses to which forest is put must be understood in specific and local terms, so too must the social implications. The Nuaulu conception of environment is not of a space in which they live but is much more like a series of fixed points to which particular clans and individuals are connected. These points are objects in an unbounded landscape linked to their appearance in myths; use of land is at every turn inseparable from specific sacred knowledge, sometimes mutually contradictory and obscure although never absent.

The undeniable effect of merging practical usefulness, mythic knowledge and identity in the construction of the category 'forest', is to give it a moral dimension. That is, there are right and wrong ways in which to engage with it which arise from the specific social histories of parts of it, but also from its intrinsic mystical properties. Forest is unpredictable, dangerous and untamed, and various attempts are made to control it. This is reflected in ritual generally, in the specific rituals conducted prior to cultivating forest, in the charms which are used to protect travellers in the forest, in the prohibitions on certain behaviours and utterances while in the forest, and in the correct disposal of its products.

The practical implications of the inter-connections between the social and the environmental can be very important, and it is often the case that subsistence practices triggered by cultural beliefs (for instance, linked to prohibitions) appear to regulate resources. It is in the context of all of this that we must understand Nuaulu ritual restrictions on harvesting certain products at particular times. Certainly, the *effect* of all of these things may well be to conserve resources and maintain biodiversity, and in particular cases people may consciously do so. Much of what appears to be 'ecological balance' amongst forest peoples is either illusory or simply a beneficial function of low population densities and benign subsistence practices. Responses by the Nuaulu to commercial logging and transmigration in the 1980s have been essentially market-driven and short term, rather than long term and homeostatic, as might be thought to be consistent with their world view. When governments and other agencies interfere and seek to introduce 'rational' measures to conserve resources, ignorant of local cultural representations of the forest, their purposes may be meaningless to local peoples as Richards (1992) has shown for the Mende living on the edges of the Gola forest in Sierra Leone. Similarly, governments, with the best of intentions, may interfere with cultural regulators (purposeful or inadvertent) which are often more sensitive, and in the long term, more effective (e.g. Morauta *et al.* 1982).

6.   Some conclusions

In conclusion we need to emphasis three features of what indigenous people know about forest and how they use it:

1.   although different groups conceptualise nature in different ways [9], these cultural constructs are only the context in which similar kinds of knowledge are pragmatically understood;

2.   we must separate out what people know and can apply, from formal, linguistic, knowledge;

3.   it should be recognised that indigenous knowledge is always situational, variable, and changing.

To elaborate upon points two and three. The second point develops the distinction between formal linguistically-encoded knowledge which passively acknowledges diversity and functionality, and substantive knowledge which is dynamically adaptive. Although formal knowledge at one level reflects a universal tendency of the human mind, it has been observed that, surprisingly, it is quantitatively less amongst food gatherers. This would appear to be related to the social, demographic and mobility characteristics of non-agricultural peoples, where knowledge is gained essentially through personal experience, not reflected in shared terminologies. On the other hand, populations less dependent on regular agriculture have a greater substantive knowledge of non-cultivated resources, even if the terms are lexically encoded.   That this disjunction exists may also be related to the vulnerability of agriculture to periodic failure due to pests and predators, and the advantage of maintaining some kind of knowledge of 'famine foods'.

Finally, and perhaps more importantly, indigenous ecological knowledge and practice must be understood contextually, beginning at the species level and working outwards: an approach which locks specific local knowledge within increasingly more general but denser culturally-relative paradigms, and which links indigenous ecological know-how to broader subsistence and social behaviour. A pharmacologist looking at ethnobotany is understandably inclined to see potential drugs, a botanist species not recorded scientifically, food scientists new foods, material scientists new materials and the 'Body Shop' some politically correct cosmetic. However, such an approach tends to *reduce* indigenous knowledge to partial unconnected bits; or, to put it another way, knowledge is transformed into information. In the process much of the value is lost. Rather than generating selected items of information in a framework determined by the quite specialised requirements of conventional biological science and taxonomy, we should be focussing on connected *systems* of local knowledge, informed by an understanding that such knowledge is dynamic. Most folk-biological knowledge differs fundamentally from conventional science in not being organised abstractly within some convenient general-purpose classification, but rather with respect to particular contexts, defined perhaps in terms of different subsistence activities. How that knowledge is apprehended by people will be determined by culturally relative coordinates of sense perception which sometimes deviate sharply from the expectations of scientifically-trained personnel; for example, the significance of olfactory and textual stimuli compared with the purely visual. Thus, much knowledge is inaccessible except via a research strategy which allows a multi-focal approach; and if investigators additionally wish to appreciate how local people make key subsistence decisions, they must attend to the categories of knowledge which are *locally applicable*.

Indigenous knowledge is also intrinsically variable though, and subject to change. We now have convincing demonstrations that knowledge may vary qualitatively and quantitatively according to crucial social factors, such as gender. We also need to take into account variation between different rainforest populations, reflecting, for example, different modes of subsistence, but it may be artificial to separate out populations on the basis of their apparent degree of interaction with forest, the extent of acculturation or integration into the market. The writer takes a broad view of what constitutes a rainforest population, including peasants as well as those more usually thought of as forest peoples. Knowledge may pass between superficially different groups which are in contact or who extract from similar biotopes. There are many observations to the effect that food-collectors (and many intermediate groups) are particularly well adapted to the rainforest and even that this lack of knowledge effectively locks many farming peoples out of the forest, who can therefore only obtain its products as dependents of forest people. In practice though the relationship between different groups is likely to be more complex, suggesting that what might be seen as discrete bodies of knowledge tied to particular social and ethnic groups might better be seen as a division of ethnobiological knowledge reflecting the specialisms of people who have long been in contact and who share a common origin. We must be wary of inventing a category of 'traditional' peoples, whose knowledge is regarded as somehow pristine and superior, however much the temptation. Also, just as knowledge is not fixed in its contemporary distribution, so it changes through time too: the result of generations of trial-and-error testing, extensive experimental evidence, enormous individual and specialist collective experience; and because new species move in and out. It is this which development consultants, conservationists and pharmaceutical firms are now taking advantage of, but they and us also need to appreciate *how* that experience has been gained.

Notes

1.   Most emphasis on the indigenous knowledge of rainforest peoples and its commercial applications has been placed on ethnopharmacology. The literature is growing at an exponential rate, but for some examples see: Arvigo and Black (1993); Balick (1990); Elisabetsky and Posey (1994); King (1994); King and Tempstra (1994); and Schultes (1994). The economic values placed on rainforest in general terms, as these are reflected in indigenous knowledge, are discussed in Godoy and Bawa (1993); Peters *et al.* (1989); and Plotkin and Famalore (1992). See also, Panayotou and Ashton (1992). On the role of such knowledge in conservation and sustainable and community development, see Martin (1994, 1995: 223-251), Williams and Baines (1993) and Posey *et al.* (1984).

2.   For the Penan case, see Ritchie (1994) and for the Yanomami, Colchester (1992). More general information is to be found in Colchester and Lohmann (1993).

3.   The broader picture is given in Flenley (1979: 77-100); for Southeast Asia, see
       Maloney (1993); Whitmore (1990: 94); Glover (1977: 160); for the Amazon,
       see Bal'ee (1993, 1994); Roosevelt (1994); and, for equatorial Africa, see
       Kadomura (1990).

4.   See references cited in note 3, and also (for Southeast Asia) Dunn (1975);
       Flenley (1979: 122); Hutterer (1983: 196); and Medway (1977).

5.   Ellen (1985b: 568); cf. Rambo (1985: 68) for *Koompassia excelsa* among the
       Semang of peninsular Malaysia.

6.   Some of the more general issues of typology are considered briefly in Ellen
       (1988, 1994).

7.   Bahuchet (1983); Peterson (1978). For selected examples from other parts of
       the world, see: Dunn (1975); Morris (1982); and Roosevelt (1994).

8.   We now have a considerable body of systematic data on the ethnobiological
       knowledge of rainforest peoples, mostly concerning plants. Major monographs
       include Bal'ee (1994); Conklin (1954); Ellen (1983a); Friedberg (1990); Revel
       (1990); and Taylor (1990). Berlin's important work on Amazonian Peru is best
       accessed through the bibliography attached to Berlin (1992). Some other
       important sources are listed in the bibliography to Brown (1985), and in
       Conklin (1972).

9.   Some of these differences are documented in the burgeoning literature on 'the
       cultural construction of nature'. Compare, say, Ingold (1996); Descola (1994);
       Ellen (1996); and Strathern (1980).

References

Alcorn, J.B. (1981). Factors influencing botanical resource perception among the Huastec: suggestions for
       future ethnobotanical enquiry, *Journal of Ethnobiology*, **1**, 221-230.

Arvigo, R. and Balick, M.J. (1993). *Rainforest Remedies: 100 Healing Herbs of Belize*, Lotus Press, Wilmot,
       Wisconsin.

Bahuchet, S. (1983). Aka-farmer relationship in central African rainforest, in R.B. Lee and E. Leacock (eds.),
       Politics and History in Band Societies, Cambridge University Press, Cambridge, pp. 189-211.

Bal'ee, W. (1993). Indigenous transformation of Amazonian forests: an example from Maranao, Brazil,
       L'Homme, **33** (2-4), 231-254.

Bal'ee, W. (1994). *Footprints of the Forest: Ka'apar Ethnobotany - the Historical Ecology of Plant
       Utilization by an Amazonian People*, Columbia University Press, New York.

Balick, M.J. (1990). Ethnobotany and the identification of therapeutic agents from the rainforest, in *Ciba
       Foundation Symposium 154, Bioactive Compounds from Plants*, Wiley, Chichester, pp. 22-39.

Barnard, A. (1983). Contemporary hunter-gatherers: current theoretical issues in ecology and social
       organisation, *Annual Review of Anthropology*, **12**, 193-214.

Berlin, B. (1992). *Ethnobiological Classification: Principles of Categorization of Plants and Animals in
       Traditional Societies*, Princeton University Press, Princeton, New Jersey.

Boster, J. (1986). Exchange of varieties and information between Aguaruna manioc cultivators, *American*

Anthropologist, **88** (2), 428-436.

Boster, J. (1994). Inferring decision making from behavior: an analysis of Aguaruna Jivaro manioc selection, *Human Ecology*, **12** (4), 347-358.

Brown, C.H. (1985). Mode of subsistence and folk biological taxonomy, *Current Anthropology*, **26** (1), 43-53.

Carneiro, R. (1978). The knowledge and use of rainforest trees by the Kuikuru Indians of central Brazil, in R.I. Ford (ed.), *The Nature and Status of Ethnobotany* (Anthropological Papers 67), Museum of Anthropology, University of Michigan, Ann Arbor, Michigan, pp. 201-216.

Carneiro, R.L. (1988). Indians of the Amazonian forest, in J.S. Denslow and C. Padoch (eds.), *People of the Tropical Rainforest*, University of California Press, Berkeley, pp. 73-86.

Colchester, M. (1992). Yanomami demand their land, *Third World Resurgence*, **6**.

Colchester, M. and Lohmann, L. (eds.), (1993). *The Struggle for Land and the Fate of the Forests*, World Rainforest Movement: Penang, Malaysia.

Conklin, H.C. (1954). *The Relation of Hanun'oo Culture to the Plant World*, PhD dissertation, Yale University, New Haven.

Conklin, H.C. (1957). *Hanun'oo Agriculture, a Report on an Integral System of Shifting Cultivation in the Philippines*, FAO, Rome.

Conklin, H.C. (1972). *Folk Classification: a Topically Arranged Bibliography of Contemporary and Background References Through 1971*, Department of Anthropology, Yale University, New Haven.

Descola, P. (1994). *In the Society of Nature: a Native Ecology in Amazonia* (Cambridge Studies in Social and Cultural Anthropology 93), Cambridge University Press, Cambridge.

Dove, M.R. (1983a). Forest preference in swidden agriculture, *Tropical Ecology*, **24** (1), 122-142.

Dove, M.R. (1983b). Theories of swidden agriculture and the political economy of ignorance, *Agroforestry Systems*, **1** (3), 85-89.

Dunn, F.L. (1975). *Rainforest Collectors and Traders: a Study of Resource Utilization in Modern and Ancient Malaya,* Monographs of the Malaysian Branch of the Royal Asiatic Society 5.

Dwyer, P.D. (1996). The invention of nature, in R.F. Ellen and K. Fukui (eds.), *Redefining Nature: Ecology, Culture and Domestication,* Berg, Oxford, pp. 157-186.

Elisabetsky, E. and Posey, D. (1994). Ethnopharmacological search for antiviral compounds: treatment of gastrointestinal disorders by Kayap'o medical specialists, in G. Prance, D. Chadwick and J. Marsh (eds.), *Ethnobotany and the Search for New Drugs* (Ciba Foundation Symposium 185), Wiley, Chichester, pp. 77-90.

Ellen, R.F. (1985a). Comment on 'Mode of subsistence and folk biological taxonomy' by Cecil H. Brown, *Current Anthropology*, **26** (1), 55-56.

Ellen, R.F. (1985b). Patterns of indigenous timber extraction from Moluccan rain forest fringes, *Journal of Biogeography*, **12**, 559-587.

Ellen, R.F. (1988). Foraging, starch extraction and the sedentary lifestyle in the lowland rainforest of central Seram, in J. Woodburn, T. Ingold and D. Riches (eds.), *History, Evolution and Social Change in Hunting and Gathering Societies*, Berg, London, pp. 117-134.

Ellen, R.F. (1993a). *Nuaulu Ethnozoology: a Systematic Inventory,* (CSAC Monographs 6), Centre for Social Anthropology and Computing, in co-operation with the Centre for South-East Asian Studies, University of Kent, Canterbury, Kent.

Ellen, R.F. (1993b). Rhetoric, practice and incentive in the face of changing times: a case study of Nuaulu attitudes to conservation and deforestation, in K. Milton (ed.), *Environmentalism: the View from Anthropology,* Routledge, London, pp. 126-143.

Ellen, R.F. (1994). Modes of subsistence: hunting and gathering to agriculture and pastoralism, in T. Ingold (ed.), *Companion Encyclopaedia of Anthropology: Humanity, Culture and Social Life*, Routledge,

London, pp. 197-225.

Ellen, R.F. (1996). The cognitive geometry of nature: a contextual approach, in P. Descola and G. Palsson (eds.), *Nature and Society: Anthropological Perspectives*, Routledge, London, pp. 103-123.

Ellen, R.F. : in press Modes of subsistence and ethnobiological knowledge: between extraction and cultivation in Southeast Asia, in D.L. Medin and S. Atran (eds.), *Ethnobiology*, M.I.T. Press.

Flenley, J.R. (1979). *The Equatorial Rain Forest: a Geological History,* Butterworths, London.

Fox, R.B. (1953). The Pinatubo Negritos: their useful plants and material culture, *Philippine Journal of Science*, **81**, 173-414.

Friedberg, C. (1990). *Le Savoir Botanique des Bunaq: Percevoir et Classer dans le Haut Lamaknen (Timor, Indonesie),* Memoires du Museum National d'Historie Naturelle, Botanique 32, Museum National d'Historie Naturelle, Paris.

Gianno, R. (1990). *Semelai Culture and Resin Technology,* Memoirs of the Connecticut Academy of Arts and Sciences No. 22, The Connecticut Academy of Arts and Sciences, New Haven.

Glover, I.C. (1977). The Hoabinhian: hunter-gatherers or early agriculturalists in South-east Asia, in J.V.S. Megaw (ed.), *Hunters, Gatherers and First Farmers Beyond Europe,* Leicester University Press, Leicester, pp. 145-166.

Godoy, R.A. and Bawa, K.S. (1993). The economic value and sustainable harvest of plants and animals from the tropical rainforest: assumptions, hypotheses and methods, *Economic Botany*, **47**, 215-219.

Hays, T.E. (1982). Utilitarian/adaptionist explanations of folk biological classification: some cautionary notes, *Journal of Ethnobiology*, **2** (1), 89-94.

Headland, T.N. (1987). The wild yam question: how well could independent hunter-gatherers live in a tropical rainforest ecosystem? *Human Ecology*, **15** (4), 463-492.

Headland, T.N. and Bailey, R.C. (1991). Introduction: have hunters and gatherers ever lived in tropical rainforest independently of agriculture? *Human Ecology*, **19** (2), 115-122.

Hoffman, C. (1984). Punan foragers in the trading networks of southeast Asia, in C. Schrire (ed.), *Past and Present in Hunter-Gatherer Studies,* Academic Press, London, pp. 123-149.

Hutterer, K.L. (1983). The natural and cultural history of southeast Asian agriculture: ecological and evolutionary considerations, *Anthropos*, **78**, 169-212.

Ichikawa, M. (1986). Ecological bases of symbiosis, territoriality and intra band cooperation of Mbuti pygmies, *Sprache und Geschichte in Afrika*, **7** (1), 161-188.

Ichikawa, M. (1992). Traditional use of tropical rainforest by the Mbuti hunter gatherers in Africa, in N. Itoigawa, Y. Sugiyama, G.P. Sackett and R.K.R. Thompson (eds.), *Topics in Primatology: Behaviour, Ecology and Conservation,* University of Tokyo Press, Tokyo, pp. 305-317.

Ingold, T. (1996). Hunting and gathering as ways of perceiving the environment, in R. Ellen and K. Fukui (eds.), *Redefining Nature: Ecology, Culture and Domestication*, Berg, Oxford, pp. 117-156.

Kadomura, H. (1990). Environmental change in tropical Africa, *Global EnvironmentalProblems*, **2**, 6-93 (in Japanese).

King, S.R. (1994). Establishing reciprocity: biodiversity, conservation and new models for cooperation between forest-dwelling people and the pharmaceutical industry, in T. Greaves (ed.), *Intellectual Property Rights for Indigenous People, a Sourcebook*, Society for Applied Anthropology, Oklahoma, pp. 69-82.

King, S.R. and Tempstra, M.S. (1994). From shaman to human clinical trials: the role of industry in ethnobotany, conservation and community reciprocity, in G. Prance, D. Chadwick and J. Marsh (eds.), *Ethnobotany and the Search for New Drugs*, Ciba Foundation Symposium No. 185, Wiley, Chichester, pp.197-206.

Maloney, B.K. (1993). Climate, man, and thirty thousand years of vegetation change in north Sumatra, *Indonesian Environmental History Newsletter*, **2**, 3-4.

Martin, G.J. (1994). Conservation and ethnobotanical exploration, in G. Prance, D. Chadwick and J. Marsh (eds.), *Ethnobotany and the Search for New Drugs*, pp. 229-239, Ciba Foundation Symposium No. 185, Wiley, Chichester.

Martin, G.J. (1995). *Ethnobotany: a methods manual*, Chapman and Hall, London.

Medway, Lord (1977). The Niah excavations and an assessment of the impact of early man on mammals in Borneo, *Asian Perspectives*, **20**, 51-69.

Morauta, L., Pernetta, J. and Heaney, W. (eds.), (1982). *Traditional Conservation in Papua New Guinea: Implication for Today*, Institute for Applied Social and Economic Research Monograph No. 16, Institute for Applied Social and Economic Research, Papua New Guinea.

Morris, B. (1982). *Forest Traders: a Socio-economic Study of the Hill Pandaram*, London School of Economics Monographs on Social Anthropology No. 55, Athlone Press, London.

Panayotou, T. and Ashton, P.S. (1992). *Not by Timber Alone: Economics and Ecology for Sustaining Tropical Forests*, Island Press, Washington DC.

Peters, C.M., Gentry, A.H. and Mendelsohn, R.O. (1989). Valuation of an Amazonian rainforest, *Nature* **339**, 665-666.

Peterson, J.T. (1978). *The Ecology of Social Boundaries: Agta Foragers of the Philippines*, Illinois Studies in Anthropology No. 11, University of Illinois Press, Urbana.

Plotkin, M.R. and Famolore, L. (eds.) (1992). *Sustainable Harvest and Marketing of Rainforest Products*, Conservation International/Island Press, Washington DC.

Posey, D. (1982). Nomadic agriculture of the Amazon, *Garden*, **6** (1), 18-24.

Posey, D. (1988). Kayap'o Indian natural-resource management, in J.S. Denslow and C. Padoch (eds.), *People of the Tropical Rainforest*, University of California Press, Berkeley, pp. 89-90.

Posey, D., Frechione, J., Eddins, J., da Silva, L.F., Myers, D., Case, D. and Macbeth, P. (1984). Ethnoecology as applied anthropology in Amazonian development, *Human Organization*, **43**, 95-106.

Rambo, A.T. (1979). Primitive man's impact on genetic resources of the Malaysian tropical rain forest, *Malaysian Applied Biology*, **8** (1), 59-65.

Rambo, A.T. (1985). *Primitive Polluters: Semang Impact on the Malaysian Tropical Rain Forest Ecosystem*, University of Michigan Museum of Anthropology Anthropological Papers No. 76, University of Michigan, Ann Arbor, Michigan.

Reichel-Dolmatoff, G. (1976). Cosmology as ecological analysis: a view from the rain forest, *Man*, n.s. **11** (3), 307-318.

Revel, N. (1990). *Fleurs de Paroles: Historie Naturelle Palawan* (vols. 1 and 2), Peeters/SELAF, Paris.

Richards, P. (1985). *Indigenous Agricultural Revolution: Ecology and Food Production in west Africa*, Hutchinson, London.

Richards, P. (1992). Saving the rain forest? Contested futures in conservation, in S. Wallman (ed.), *Contemporary Futures: Perspectives from Social Anthropology,* Association of Social Anthropologists Monographs No. 30, Routledge, London, pp. 138-153.

Ritchie, J. (1994). *Bruno Manser: the Inside Story*, Summer Times Publishing, Singapore.

Roosevelt, A. (ed.) (1994). *Amazonian Indians from Prehistory to the Present: Anthropological Perspectives*, University of Arizona Press, Tucson.

Schultes, R.E. (1994). Amazonian ethnobotany and the search for new drugs, in G. Prance, D. Chadwick and J. Marsh (eds.), *Ethnobotany and the Search for New Drugs*, Ciba Foundation Symposium No. 185, Wiley, Chichester, pp. 106-112.

Sillitoe, P. (1991). Worms that bite and other aspects of Wola soil lore, in A. Pawley (ed.), *Man and a Half: Essays in Pacific Anthropology and Ethnobiology Honour of Ralph Bulmer*, Polynesian Society Memoir No. 48, The Polynesian Society, Auckland, New Zealand, pp. 152-163.

Strathern, M. (1980). No nature, no culture: the Hagen case, in C. MacCormack and M. Strathern (eds.),

Nature, Culture and Gender, Cambridge University Press, Cambridge, pp. 174-222.

Taylor, P.M. (1990). *The Folk Biology of the Tobelo People: A Study in Folk Classification,* Smithsonian Contributions to Anthropology No. 34, Smithsonian Institution Press, Washington DC.

Whitmore, T.C. (1990). *An Introduction to Tropical Rain Forests*, Clarendon Press, Oxford.

Williams, N. and Baines, G. (eds.), (1993). *Traditional Ecological Knowledge: Wisdom for Sustainable Development*, Centre for Resource and Environmental Studies, Australian National University, Canberra.

Yen, D. (1989). The domestication of environment, in D. Harris and G.C. Hillman (eds.), *Foraging and Farming: the Evolution of Plant Exploitation*, Unwin Hyman, London, pp. 55-75.

Professor Roy F. Ellen, Department of Sociology and Social Anthropology, Eliot College, University of Kent at Canterbury, Canterbury, Kent CT2 7NS, England

6. HISTORY AND DESTINY OF MIDDLE AMERICAN FORESTS: THE INHERITORS OF THE MAYAN LANDSCAPE

Peter A. Furley

1. Introduction

Middle America, comprising the mainland isthmus from Mexico to Panama together with the islands of the Caribbean, possesses a remarkable diversity of landscapes, mirrored in exceptional biological diversity. Across this highly dynamic stage an historical pageant that has fascinated generations of observers, fashioned and redistributed the natural resources in the unique manner that has endowed so much to the character of the region today. In particular, the forests that now occupy the Mayan lowlands provide a forceful illustration of the interactive flux between people and their resources over time.

The purpose of this account is to show how the *Selva Maya*, the forests of the Mayan region, have fared in comparison with other land cover changes, and to focus on a number of the positive as well as negative trends that prevail at the end of the current millennium.

2. The origin and nature of the vegetation

Middle America has witnessed a history of environmental turbulence affecting the present day landscapes since well before the Quaternary. Land bridges and 'stepping stone' connections have influenced plant and animal dispersal since the Cretaceous (Coney 1982). Nevertheless it is likely that a continuous land bridge did not exist until the volcanic islands making up the present day Cordillera were fused together in the late Tertiary (Murphy and Lugo 1995). Whereas there had been opportunity for dispersal across the segments of Central America for over 60m years, the Caribbean islands have been isolated since the Cretaceous and it can be assumed that their flora and fauna were derived from dispersal over sea and subsequent differentiation.

The dynamic shifts of sea level and land bridges or island connections, of climatic swings between wet and dry, of volcanic eruption and tectonic flexing and faulting, have all contributed to the patchwork nature (Figure 1) of the contemporary plant cover (Weyl 1980; Graham 1992; Graham and Dilcher 1995). Over more recent geological time, the most significant changes in the vegetation began to occur in the Pliocene (*c*. 4m yr B.P.) as cooler and drier conditions prevailed. In the Pleistocene, forests expanded during humid phases and survived in altered or restricted form during dry phases (Whitmore and Prance 1987). These confined distributions or *refugia* have engendered much controversy, although it is the locations rather than the principle which is attacked. Toledo (1982) focussed on five Pleistocene *refugia* in Mexico and Central America, which he derived (Figure 2) from analyses of endemic species. He took into account abnormal distributions of temperate and dry elements in the rain forest, drought tolerance features, and movements of rain forest species into cooler climatic zones as well as the distribution patterns of dominant species. However, not all of the putative endemic *refugia* have lived up to their prediction. For example, the gallery forests of the Maya Mountains in Belize, believed to have developed over an ancient land surface remaining as an island largely above the fluctuating seas of the

B.K. Maloney (ed.), Human Activities and the Tropical Rainforest, 101-132.

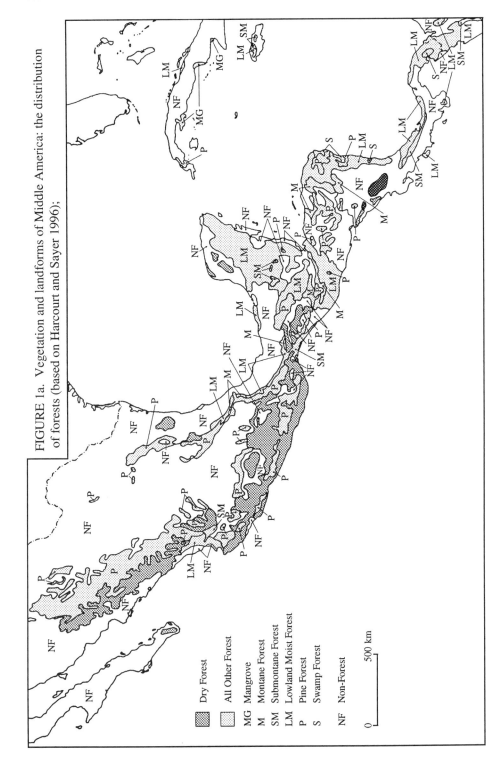

FIGURE 1a. Vegetation and landforms of Middle America: the distribution of forests (based on Harcourt and Sayer 1996);

Dry Forest
All Other Forest

MG Mangrove
M   Montane Forest
SM Submontane Forest
LM Lowland Moist Forest
P   Pine Forest
S   Swamp Forest

NF   Non-Forest

0   500 km

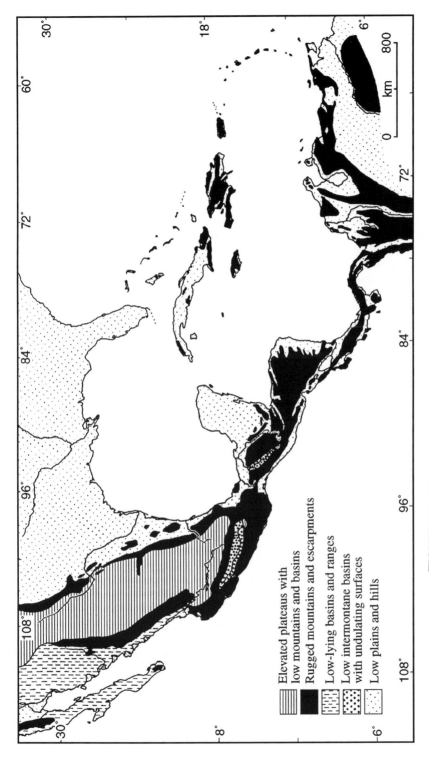

FIGURE 1b. Physiographic regions (based on West and Augelli 1986).

Mesozoic period, were thought to be potential relict areas of vegetation (Wright *et al.* 1959; Toledo 1982). Although relatively rich in species and capable of acting as sources for forest resurgence, they have not fulfilled the expectation of significant endemism (Kellman *et al.* 1996; Meave and Kellman 1994). In much more recent time the region has been buffetted by major natural events such as hurricanes (according to Neumann (1977) over 761 tropical storms were reported between 1885-1977), volcanic eruptions, earthquakes and devastating fires, all of which contribute to the continual renewal and change characteristic of the area.

Some 30 to 40 years ago, it was generally believed that the humid lowlands of Central America were covered by tropical moist forest that had been virtually unaffected by the glacial fluctuations of the northern hemisphere and relatively undisturbed by indigenous occupation. Both concepts have been shown to be completely incorrect. During the periods of ice expansion and contraction during the Pleistocene, the upland areas of Middle America witnessed marked altitudinal shifts in forest zonation (Flenley 1979; 1992), whilst the lowlands appear to have been drier with moist, dense forest being probably replaced by more open and semi-deciduous forest or arboreal savanna. Goldammer (1992: 3) stated that a generally cooler and drier tropical climate forced the pre-glacial rain forest biota to retreat to *refugia* in which the unfavourable climatic conditions could be compensated by topographic, hydrological and micro-climatic particularities of the terrain.

Finally, what was thought to be undisturbed forest has been uncovered to reveal evidence of fields, settlements and ceremonial centres of great complexity and abundance. As various groups achieved temporary ascendancy, so conversely the forests were reduced or increasingly disturbed.

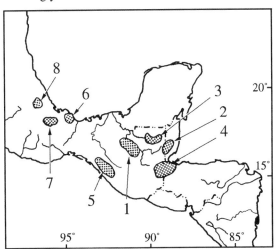

FIGURE 2. Pleistocene refugia in Central America. Primary refugia are proposed in the Lacandon area, Chiapas, Mexico (1), the Toldeo District of the Maya Mountains, Belize (2), the northern Peten, Guatemala, stretching into Cayo District, Belize (3), Lake Izabal area, Guatemala (4), and Soconusco, Chiapas, Mexico (5). Secondary refugia have been identified at Los Tuxtlas (6), Sierra de Juarez (7) and Cordoba, Mexico (9) (from Toledo 1982).

3 .   Patterns of vegetation and the nature of the forest cover

The varied environmental zones characteristic of Middle America have resulted in a diverse pattern of vegetation. The well known life-zone model (Figure 3) of Holdridge (1967, 1971) takes into account the physical environment and its biological response, and can be simplified into biotic provinces. A more pragmatic classification relevant at a local scale can be made on the basis of altitude, response to length of dry season and the permanence of water.

Taking the vertical zonation (Figure 4) first of all, the contemporary vegetation ranges from upper montane or alpine grasslands found at heights over 3500m at the upper limits of plant growth (*tierra helada*), to montane vegetation with shrubs and occasional trees (*tierra helada* and *tierra fria*), lower montane formations (*tierra fria*) which may consist of cloud forest and mixed temperate forests (*tierra templada*), and, finally, descending to lowland tropical moist evergreen and semi-evergreen forests (*tierra caliente*). The lapse rate of temperature decreases with elevation in the order of 6°C per 100m. In simplified terms, lowlands range from 0 to 500m in elevation, low mountains and hills from 500 to 2000m and montane conditions and high mountains occur over 2000m .

Within these altitudinal zones there are wetter or drier tracts, reflecting drainage and geology, which result in swamps and marshes or dry forest and thorn scrub in more permanently wet or dry conditions respectively. More saline conditions occur along the coastal plains and around inland lagoons. Mangrove forests and other forested wetlands are well developed along the flatter littorals of the Atlantic coast (Lugo 1990) and can extend well inland, probably as relict formations (Furley and Ratter 1992).

The length of the dry season responds to the broad climatic patterns of the region and to topography. In Central America, the Atlantic coasts tend to be wetter than the Pacific, reacting to the prevailing trade winds, and the north and west are much drier than the south and east. The differences are vividly expressed in many of the Caribbean islands with extremely wet windward and very dry leeward slopes. Major orographic obstacles frequently result in increased fog and cloud, giving rise to cloud forests on the windward slopes. Seasonality is a major determinant of forest type. More typical moist evergreen forest is generally found below 1000m where the dry season is relatively short (1-2 months). Increased length and intensity of the dry periods results progressively in upper canopy deciduousness, eventually leading to different forms of dry forest (Beard 1944, 1955).

In addition to the variations occasioned by aspect, altitude, rainfall and drainage, there are significant local changes caused by edaphic differences. Dry, coarse textured soils of low nutrient status may give rise to pine forests and arboreal savannas (Perry 1991), whilst patches of more mesotrophic soil can often trigger a deciduous response in areas which are predominantly savanna or evergreen forest.

Four forest formations are outstanding in Central America. Montane and upland forests have a typical pine component and run down the axis of the Central Cordillera, moist evergreen forests predominate on the Atlantic side and towards the southern junction with South America, dry deciduous forest are, conversely, found to the north, particularly in Mexico, and finally the mangrove and wetland forests are found

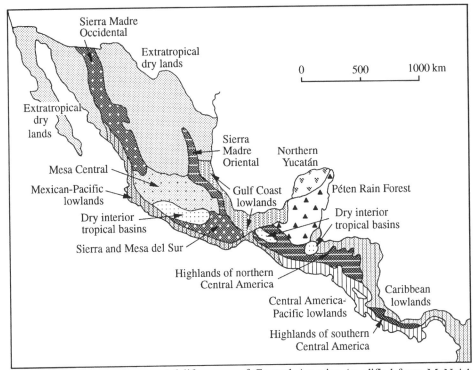

FIGURE 3.   The environmental life zones of Central America (modified from McNeish 1992).

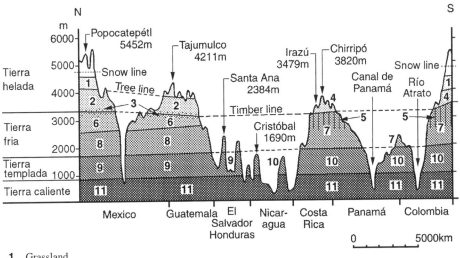

1	Grassland		
2	Grassland with pine and fir	7	Tropical cloud forest
3	Ericaceous belt	8	Mixed tropical forest with boreal elements (pine, oak, fir)
4	Tropical paramo	9	Mixed montane pine/oak forest
5	Bamboo belt	10	Tropical evergreen and semi-evergreen montane forest
6	Mixed upland boreal forest (pine, oak, fir, alder)	11	Lowland tropical evergreen and semi-evergreen forest

FIGURE 4.   Contemporary altitudinal variations in vegetation through Central America (based on Lauer (1968) and Stadmuller 1987).

on both Atlantic and Pacific shorelines, being most distinctive on flat, poorly drained and emergent coasts (see Figure 1).

3.1. MONTANE AND UPLAND FORESTS

As indicated earlier, the marked changes in forest structure and composition that accompany increasing height above sea level do not simply result from altitude but reflect exposure to prevalent winds. This is clearly illustrated in some of the Caribbean islands such as the Lesser Antilles (Figure 5). Three zones can be recognised following the work of Lugo (1981). Lower montane rain forest occurs on slopes 60 to 500m asl. In favourable, less exposed areas the forest can be closed and dense but with variable height (typically 20-30m). Where conditions are more open there is often an understorey of around 12m in height and with a distinct shrub and ground layer. Disturbance tends to lead to a greater predominance of tree ferns and *Miconia* sp. Montane thickets and palm brakes are found over ridge tops and at heights of 300-600m asl and there is frequently a dense canopy of slender trees with tall crowns at 12 to 20m. Although the woody understorey is sparse, there are many mosses and epiphytes. Palms can form a sub-climax giving rise to a patchy forest with a large proportion of *Euterpe* spp. There is little in the way of a shrub layer but a rich herbaceous layer occurs. Finally at heights over around 600m, elfin forest tends to develop in conditions of constant exposure and wetness. This forest is a single woody structure, of low stature (3-6m), gnarled, and densely covered by epiphytes.

Pine forests and pine savannas are characteristic of upland areas in much of the region (Perry 1991). These are widespread from the US border with Mexico to the south of Nicaragua. Murov (1967) regarded Central America as a secondary centre of evolution for the genus *Pinus* where interspecific hybridization took place, and there are a number of rare and endangered species, especially in Mexico.

High mountains tend to be the home for some of the greatest concentrations of endemic species in the region (Vuilleumier and Monasterio 1986). For example, 70% of the vascular flora of high mountains in Guatemala and Mexico is reportedly endemic (D'Arcy 1977). Many of these mountains attain great heights; Cotlaltepetl in Mexico reaches to around 5500m, the highest peak south of Alaska.

3.2. MOIST EVERGREEN FOREST

The most structurally and taxonomically diverse vegetation tends to be associated with the wettest zones and the centres of diversity have been discussed by Davis *et al.* (quoted in Harcourt and Sayer 1996). Even within the moist forest a broad rainfall gradient can be discerned and the wetter parts of the range may contain double the number of species of the drier forests.

A brief comparison of the climatic and vegetation maps of Middle America reveals a north-south trend. The dry forest and scrubland in the north of Mexico gives way to increasingly humid and forested landscapes towards Panama. This can be seen at a more local scale in the Yucatan Peninsula, where there is a perceptible change from scrub forest on the northern coast to increasingly moist forest southwards into Belize and Guatemala. Generally the levels of both endemism and species richness are less than in the widespread moist forests of South America, but cloud forests contain

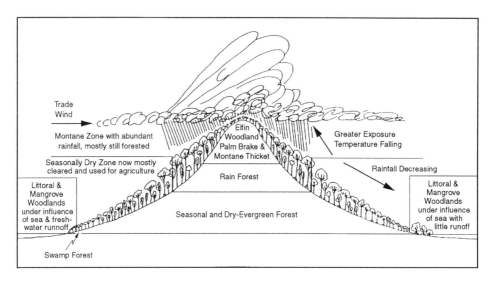

FIGURE 5. The influence of altitude and aspect on vegetation across a Caribbean island, Lesser Antilles (based on Lugo *et al.* 1981).

exceptional concentrations of endemics and are more important to the south of the isthmus, e.g. in Costa Rica.

Mexico provides a good example of the moist forest habitat (FAO 1993). Trees are mostly evergreen at rainfall levels over 2000mm/yr, particularly on the wetter Atlantic side. Where the effect of seasonality is diminished, similar forests can occur at lower rainfalls. Most of the evergreen forests are found at low altitudes on volcanic and sedimentary parent materials. The tallest trees reach heights of over 60m and attain buttress diameters greater than 60cm, although most of the trees have straight trunks with few low branches. Whilst there are numerous useful species, only about eight are of well known commercial value (Harcourt and Sayer 1996). At lower rainfalls, typically 100 to 1300mm, semi-evergreen forests are common and, although confined to the Atlantic coasts, they are the most widespread moist forest formation in Mexico. They possess sub-perennial leaves and occur where there is a dry season of around three months duration, during which time 25 to 50% of species shed all or much of their foliage. As the dry season gets longer, the forest changes to a more semi-deciduous form where a larger proportion (perhaps 50 to 75%) of the dominant trees lose their leaves.

3.3.   DRY DECIDUOUS FORESTS

Many researchers regard these as amongst the most endangered of all tropical forests (Janzen 1988; Ratter 1992) and they are believed to have been utilised for agriculture for over 5000 years (Bullock *et al.* 1995) and to have been disturbed for much longer (*c.* 11,000 years in Panama, according to Bush *et al.* 1992).

Murphy and Lugo (1995) recognise 28 tropical and sub-tropical forest life zones in Central America and the Caribbean, and they identify something like 50% of the region as lying within the dry forest zone (*sensu* Holdridge 1967). Deciduousness results from dryness and a high coefficient of rainfall variability, which may directly reflect seasonality, and be accentuated by porous substrates (sands and gravels) or may be associated with nutrient rich conditions in more typically dystrophic areas. In Central America the main determinant is climate: 'most ecosystems of the tropical and subtropical latitudes are seasonally stressed by drought' (Mooney *et al.* 1995). Increases in tree basal area have been shown in the Yucatan, for example, to be directly related to the amount of rainfall over the previous two or more years (Whigham *et al.* 1990). They are mostly found to the Pacific side in a narrow continuous band (Figure 1) from Mexico to Coast Rica, with outliers in eastern Mexico and have been extensively cut and degraded nearly everywhere, processes which augmented by the effects of fire.

The dry forests are not considered to be as rich floristically as moist evergreen forests but they are visually and phytosociologically distinctive. Frequently the associated wildlife is actually greater than in moist forest. The dry deciduous forests are extremely important in terms of endemic species, even if the figures for plant diversity are less than in comparable South American forests such as the *chaco-cerrado-caatinga* belt stretching from northern Paraguay and Argentina through central to the north east of Brazil. The Costa Rican forests are amongst the best described. Similar sites in terms of rainfall have been shown to result in very different plant communities as a product of varied soils and incidences of flooding. Gentry (1995) claims that dry forests are impoverished sub-sets of moist forest, but this is arguable since there is little overlap in tree species and they seem to have developed an individuality over a long period of time.

3.4.   MANGROVE FORESTS AND ASSOCIATED WETLANDS

Mangroves are pan-tropical formations and represent a varied group of woody trees and shrubs adapted to brackish or salt water. True mangroves, according to Tomlinson (1986) fulfil a number of criteria including morphological and physiological adaptations to the environment and taxonomic distinction from terrestrial associates. In the wider sense, which is more relevant to the practical conditions found in Middle America, the term mangrove ecosystem can be used to cover both the true mangroves and the related arboreal wetlands and can be extended to include the non-arboreal wetlands, all of which form an indistinguishable continuum in many locations.

Mangroves have spread into five biogeographic zones throughout the world (Saenger *et al.* 1983), with the Caribbean forming a sub-group within the Americas (Chapman 1977). The trees of the Caribbean group are floristically poor, comprising red, black and white mangrove (*Rhizophora mangle, Avicennia germinans, Laguncularia racemosa*) and an associated buttonwood (*Conocarpus erectus*) although there are numerous attached ground and epiphytic plants.

Mangroves fulfil a host of roles and yet have often been both ignored and abused as a resource. Not only are they the home to a rich wildlife with high biological productivity, but they are also nurseries for fish, crustaceans and molluscs which become more visible and economically perceptible in offshore coral reefs and in the fishing industry. Mangrove coastlines also act as filters for sediment and pollution and frequently provide important protective barriers against storms and coastal erosion. They are resilient and dynamic and colonise emergent and newly deposited substrates. By virtue of its size, Mexico possesses the largest area of mangrove but extensive stretches occur along most of the coastlines, notably on the Atlantic side, and on most of the islands, particularly Cuba. Much of this is under attack at present, for farming, construction or as part of the burgeoning tourist industry in the region. The figure for Mexico has dropped from 15,000 km^2 in the 1970s to around 6000 km^2 or less in 1990 (Harcourt and Sayer 1996). Costa Rica and Guatemala provide examples of countries with mainly Pacific mangrove resources. The Costa Rican Pacific coast is supposedly 35% mangrove (WRI 1995), but Leonard reported (1987) that, by the late 1970s, 40% of the land had been cleared for shrimp ponds, salt pans and other forms of coastal development.

To summarise, the distribution pattern of these forests is closely related to issues of biodiversity and conservation. Two major problems have been identified by Gentry (1992, 1995), those of diversity and endemism. In Central America it has been shown that diversity tends to increase southwards through the isthmus, associated with increasing rainfall and the incidence of moist evergreen forest. However, at a more detailed scale, within-community (alpha) diversity can vary dramatically from place to place. Endemism is only partly related to diversity and the highest levels seem to occur in isolated patches, especially in cloud forests, highly dissected montane areas and on islands.

4.   Disturbance and deforestation

Despite a long history of exploitation, much of the forest had regenerated in the lowlands of Middle America after abandonment by the Maya and before the Spanish Conquest and European occupation. The current forest cover therefore reflects not only the traditional areas of settlement over historic time but also present day location and access. Large areas of moist, dry and wetland forest remain, mostly associated with less developed and populated areas, although possibly as much as 2/3rds of the original rain forests of Central America have been lost (Nations and Komer 1983). Predictably the centres of population pressure such as the southern central plateau of Mexico, the Central American spine of the Cordilleras, especially densely populated El Salvador and the central uplands of Costa Rica, correspond to the zones of greatest deforestation and disturbance (Rudel and Roper 1996).

The *Selva Maya* still remains one of the most significant forest blocks in Middle America but the past 20 or 30 years have witnessed an accelerating pace of devastating destruction. With Southeast Asia, Central America and Mexico have the unwelcome distinction of the highest regional deforestation rates in the world, around twice the average for the tropics (WRI 1995). The high rates, consistent over the 1980s and continued into the present decade, have been attributed to the clearing of small tracts of remaining intact forest in areas of extreme population pressure (FAO 1993). It is nevertheless true that the broad generalisations inevitably miss much of the subtlety and significance of detailed analysis, for example information on disturbance beneath the forest canopy or clearings of less than the 10% used to define deforestation is mostly lacking.

The broad pattern of forest change over the decade 1981-90 is presented in Table 1. Annual deforestation rates can be seen to be over 1% for most of the states within Central America and the Caribbean and are very high even in conservation-minded countries such as Costa Rica. Jamaican forests are under particular threat and many of the Caribbean islands are desperately degraded (for example the relatively low figures for Haiti do not reveal the already high level of forest clearance). Of particular significance to conservationists is the recent trend to cut 'primary' forest (as defined by the FAO). The figures for Mexico, Nicaragua, Guatemala and Panama shown in Table 1 suggest a growing pressure on the remaining moist forests, which have hitherto been less accessible. Mexico is believed to have possessed 29m ha of tropical forest, which had fallen to 20m in 1950 and to 11.4m in 1986, partly as a result of government colonisation and land settlement schemes (FAO 1993). The picture of replanting and secondary forest growth is depressing, for although there have been global, regional and national measures to protect important areas, initiatives for restoring degraded forest or reafforestation for essential timber and fuelwood needs, are hardly evident at all.

An interesting innovation in the latest FAO world estimates for forest loss, is the breakdown into individual ecosystems. This affords a much more detailed view of the total forest picture across the region. Hill and montane forest is defined as all forms of tree cover existing at elevations over 800m. In addition to rain forest, the moist deciduous forests emerge as a highly significant component. Dry deciduous forests are only important within Mexico, where they are closely related to climate, and in the Bahamas and northern Yucatan, where they are (Table 2) also related to low elevation (reduced rainfall), together with calcareous parent soils.

There is little reason to suppose that the figures for the 1980s will have changed in any meaningful way in the 1990s and this reinforces the conclusion that most of Middle America is seriously deforested. At the same time there are positive signs; a number of factors are slowing down the pace of clearance and increasing the area under protection. Of particular interest in this context is the great lowland forest occupied by the Maya in the Yucatan Peninsula.

5.   The Mayan Forests

The forests extending north of the central Cordillera towards the Yucatan have witnessed one of the most fascinating histories of any tropical forest. The character of these forests changes from moist evergreen forms abutting onto the mountains of Chiapas in Mexico including the well known Mayan Lacandon, and the traditional Maya Highlands in Guatemala and to a lesser extent in Honduras. The forests become progressively drier northwards with moist deciduous forests in Guatemalan Peten and Belize and eventually dry forest and thorn scrub at the northernmost part of Yucatan State. Despite incursions along major roads and constant reductions and disturbance at the edges, notably in Mexico, the forest persists today as one of the most extensive and valuable tracts in Middle America. Its abundant flora and fauna, representing the interchange

	Extent of forest and wooded land (000) ha				Annual deforestation				Annual logging of closed broadleaf forest, 1981-90			Plantations (000 ha)	
	1990 Natural forest	1990 Other wooded land	1980 Natural forest	1980 Other wooded land	Total forest 1981-90 Extent (000 ha)	Total forest 1981-90 %	1981-85 Extent (000 ha)	1981-85 %	Extent (000 ha)	% of closed forest	% that is primary forest	Extent 1990	Annual change 1981-90
Temperate North America	456737	292552	X	X	X	X	X	X	X	X	X	X	X
Canada	247164	206136	299154	X	X	X	X	X	X	X	X	X	231
United States	209573	86416	299154	X	317	0.1	X	X	X	X	X	X	X
Central America and Mexico	68096	X	79216	X	1112	1.4	1022		90	0.4	65	273	17
Costa Rica	1428	X	1923	X	50	2.6	65		34	2.6	27	40	4
El Salvador	123	X	155	X	3	2.1	5		X	X	X	6	1
Guatemala	4225	X	5038	X	81	1.6	90		3	0.1	50	40	3
Honduras	4605	X	5720	X	112	1.9	90		2	0.1	19	4	0
Mexico	48586	X	55366	X	678	1.2	615		4	0	94	155	8
Nicaragua	6013	X	7254	X	124	1.7	121		45	0.9	92	20	2
Panama	3117	X	3761	X	64	1.7	36		3	0.1	71	9	1
Caribbean Subregion	47115	X	48333	X	122	0.3	26		42	0.1	73	442	23
Belize	1996	X	2046	X	5	0.2	9		3	0.2	5	3	0
Cuba	1715	X	1888	X	17	0.9	2		3	0.2	8	350	19
Dominican Republic	1077	X	1428	X	35	2.5	4		0	0	61	10	0
Guyana	18416	X	18597	X	18	0.1	3		9	0	91	12	1
Haiti	23	X	38	X	2	3.9	2		1	7.7	11	12	1
Jamaica	239	X	507	X	27	5.3	2		11	0.4	91	21	1
Suriname	14768	X	14895	X	13	0.1	3		3	0.1	94	12	0
Trinidad and Tobago	155	X	192	X	4	1.9	1		1	1.8	4	18	0

From WRI 1994-95, based on FAO and EC data

TABLE 1. Forest resources.

	Total forest		Rainforest		Moist Deciduous Forest		Hill and Montane Forest		Dry Deciduous Forest		Very Dry Forest		Desert	
	1990 extent (000 ha)	Percent annual change 1981-90	1990 extent (000 ha)	Percent annual change 1981-90	1990 extent (000 ha)	Percent annual change 1981-90	1990 extent (000 ha)	Percent annual change 1981-90	1990 extent (000 ha)	Percent annual change 1981-90	1990 extent (000 ha)	Percent annual change 1981-90	1990 extent (000 ha)	Percent annual change 1981-90
Central America and Mexico	68096	1.4	11154	1.8	12267	1.5	36296	1.4	1590	1.6	759	1.8	1424	2
Costa Rica	1428	2.6	625	2.6	0	0	802	2.6	0	0	0	0	0	0
El Salvador	123	2	33	2	12	2	79	2	0	0	0	0	0	0
Guatemala	4225	1.6	2542	1.6	731	0	953	2.5	0	0	0	0	0	0
Honduras	4605	2	1286	2	437	2	2882	1.9	0	0	0	0	0	0
Mexico	48586	1.2	2441	1	11110	1.5	31261	1.1	1590	1.6	759	1.8	1424	2
Nicaragua	6013	1.7	3712	1.7	348	1.7	1953	1.7	0	0	0	0	0	0
Panama	3117	1.7	1802	1.6	67	0.1	1249	2	0	0	0	0	0	0
Caribbean Subregion	47115	0.3	31428	0.1	12991	0.5	2639	0.8	49	2	5	1.7	4	2
Antigua and Barbuda	10	0	0	0	10	0	0	0	0	0	0	0	0	0
Bahamas	186	1.9	0	0	124	1.9	6	2.5	47	1.9	5	1.7	4	1.7
Belize	1996	0.2	1741	0	238	0	16	0	0	0	0	0	0	0
Cuba	1715	0.9	114	0.9	1247	0.9	352	0.9	2	0	0	0	0	0
Dominica	44	0.6	44	0.6	0	0	0	0	0	0	0	0	0	0
Dominican Republic	1077	2.5	341	2.5	273	2.5	463	2.5	0	0	0	0	0	0
French Guiana	7997	0	7993	0	3	0	0	0	0	0	0	0	0	0
Grenada	6	5	0	0.3	6	0	0	0	0	0	0	0	0	0
Guadeloupe	93	0.3	93	0	0	0	0	0	0	0	0	0	0	0
Guyana	18416	0.1	11671	0	5078	0.3	1668	0.1	0	0	0	0	0	0
Haiti	23	3.9	5	3.8	9	4	10	3.8	0	0	0	0	0	0
Jamaica	239	5.3	122	5.3	113	5.3	3	5.7	0	0	0	0	0	0
Martinique	43	0.4	43	0.4	0	0	0	0	0	0	0	0	0	0
Puerto Rico	321	1.5	49	1.7	151	1.5	121	1.5	0	0	0	0	0	0
Saint Kitts and Nevis	13	0	0	0	13	0	0	0	0	0	0	0	0	0
Saint Lucia	5	3.8	5	3.8	0	0	0	0	0	0	0	0	0	0
Saint Vincent	11	2.1	10	1.7	0	0	0	0	0	0	0	0	0	0
Suriname	14768	0.1	9042	0	5726	0.2	0	0	0	0	0	0	0	0
Trinidad and Tobago	155	1.9	155	1.9	0	0	0	0	0	0	0	0	0	0

From WRI, based on FAO data

TABLE 2. Tropical forest losses.

between the neotropical and nearctic biotas, allied to its extraordinary archaeological richness, have generated calls for different zones to be classified as Biosphere Reserves and for the major block extending across northern Peten into Mexico and Belize as a World Heritage site. If the forests cannot be preserved in these areas, there is little hope for less favoured remnants.

The chronology of forest occupation and abandonment is by no means certain but several generally agreed stages may be recognised. The initial point is the great age of human resource exploitation in the area. Identifiable Archaic sequences are claimed to go back to 9000-7500 B.C. in Belize, parts of Chiapas, and Tehuacan in Mexico (MacNeish 1992). The earliest inhabitants appear to have been hunter-collectors and artifacts have been found in the more open, lowland pine savannas. Other cultural stages have been adduced leading to the arrival of corn-based agriculture (probably around 3000-2000 B.C.). The first corn cultivation in Belize has been identified at Pulltrouser Swamp and dates from around 2500 B.C. MacNeish (1992) proffers a sequence of development from roving bands of Hunter Collectors to more affluent Foraging Bands to more settled Foraging Villages, Horticultural and, finally, semi-permanent Agricultural Villages. In whichever way the settlement of the region evolved, the initial penetration into the forest is likely to have been riverine and through lagoons and waterways within the wetlands (Hammond 1978).

Increasing activity gradually supported more complex societies and the emergence of stone built cities and ceremonial centres. At their peak, varying across five centuries, the individual 'city states' seem to have occupied most of the region now covered by forest, developing sophisticated agricultural techniques to cope with the increasing population (see Maloney, this volume). It is estimated that a very large proportion of the original forest must have been cut and possibly cleared several times over the period of the Maya occupation although not necessarily synchronously. An example of the clearance of forest, including pine for fuelwood, is illustrated at Copan in Honduras (Abrams and Rue 1988).

The Maya may have eventually over-exploited their forest resources, but their land husbandry techniques still offer many ideas for the potential management of the area. They introduced a series of diverse systems to cope with the complex pattern of local environments. These environments included well watered river valleys and depressions (*bajos*), rich but variable and often fragile soils [1], seasonally flooded wetlands which are increasingly saline towards the coast, as well as low sloping limestone plains and hills. Although there undoubtedly always was a rich diversity of sites, the precise configuration of wetland, swamp, dry knolls and firm ground may not always have been constant (Wiseman 1978), especially along the emergent coastline. Land management depended upon a close association of agricultural and forest techniques (Gliessman 1990). They included a number of characteristic features, for example:

    i. hydraulic systems for storing, regulating and distributing water (see for example, Turner and Harrison 1983; Turner and Denevan 1989; Gomez-Pompa 1991).

    ii. raised field and drained field systems (*chinampas*) (Gleissman 1990; Siemens and Puleston 1972; Darch 1983).

iii. use of terraces and rotational practices, usually in conjunction with *milpa* farming (Lundell 1940; Turner 1983).

iv. forest gardens (present today, and assumed to be widespread in the past) (Wiseman 1978), with raised beds for seed germination (*caanches*) and/or tree planting for fruit, shade, fuelwood, seeds (Figure 6a).

v. use of successional stages of natural forest growth and re-growth (Figure 6b).

vi. conservation of forest patches for specific purposes with planting and removal of selected species (*Pet Kot*, cf. Gomez-Pompa *et al.* 1987; Gomez-Pompa 1991).

A great diversity of plants was utilised (Sosa *et al.* 1985a,b) and the indigenous inhabitants of the region possessed a detailed knowledge of the flora and fauna (Hartig 1979). Of particular interest is their ability to recognise and exploit the regrowth stages of forest or to reconstruct an 'artificial rain forest' or forest garden. It has further been suggested that they had acquired a good sense of soil variability and potential for use (Beltran 1959; Flores and Ucan 1983). The influence of slope and drainage along with associated soil properties also guided the form of forest management. Wiseman (1978) has postulated differing management techniques along two edaphic gradients in his study of the central lakes region of Guatemalan Peten (Figure 7). The wetlands and swamps also provided a different resource base and challenge for utilisation. A combination of raised fields in wetter tracts and drained fields in the seasonally wet areas proliferated as a system and is well developed in a number of localities (for example in Quintana Roo in Mexico or northern Belize (Furley *et al.* 1995).

Increasingly, evidence is being uncovered which illustrates how the Mayan agro-silvicultural techniques were developed (Barrera *et al.* 1977). The immense variety of food producing systems over time and space has been noted for many years (Harrison 1978). Forest management consisted of various activities to select, cultivate, protect and introduce trees in shifting cultivation plots (*milpas*) or fallow land, coppicing of numerous species (by cutting down to *c.* 50cm above the ground and encouraging multi-bole regrowth), 'plantations' of favoured species (possibly cacao for the elite) within the natural forest and deliberate plantings around houses and urban centres, as living fences, and along trails (Gomez-Pompa 1987). Religion may also have played a role in the preservation of certain species, such as the cotton tree (*Ceiba pentandra*), which is still revered today, as individuals or in sacred groves.

The Maya utilised a rich array of natural resources and the full mosaic of existing ecosystems including freshwater and marine resources. In order to achieve a sustained output over many centuries, they must have acquired an advanced and sophisticated level of management. Whatever the causes of the eventual decline in the Maya civilization it is unlikely, as Gomez-Pompa (1987) has suggested, that it was a result of poor forest management techniques, although it may have resulted from excess pressure on forest resources for agricultural land and fuelwood as well as extractive products and timber.

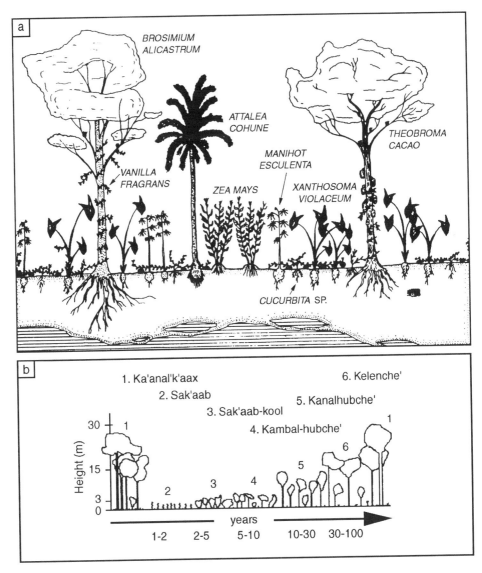

FIGURE 6. Mayan land husbandry (a) spatial distribution of crops, shrubs and trees in an 'artificial forest' or forest garden (from Wiseman 1978); (b) vegetation classification based on successional growth stages (from Gomez-Pompa 1991).

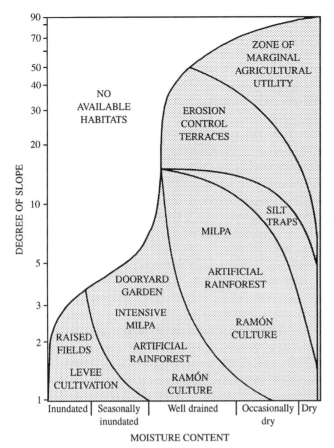

FIGURE 7. Mayan agricultural systems according to slope angle and soil moisture content (after Wiseman 1978).

6. Lessons from the past and directions for the future: forests and forest management in Belize

Belize is worth examining in greater detail as a case study of the wider Mayan realm because, despite its small size (22,963 km² and a population of around 200,000), it possesses some of the most extensive stretches of varied forest in Central America (Figure 8), and one of the highest proportions of conserved land per inhabitant. The 16 reserves contain around 20% of the country's forests, with 73% of the pine and 25% of the broadleaf resources (Harcourt and Sayer 1996), and with some 4000 species of flowering plants, including 700 trees.

This has come about partly by historic accident and partly by design. The colonial history of Belize until the global improvement in air transport in the 1960s and 1970s and the impetus of independence in 1981, was largely one of lack of development. This had the unforseen benefit that the Belizean government inherited considerable tracts of national land which were relatively undisturbed in comparison to those of its more populous neighbours. Even the privately owned forests were not logged destructively in spite of the long colonial history of exploitation, initially of rosewoods and dyewoods and later, over the past 150 years, of hardwoods, such as mahogany and cedar, and softwoods like pine. Some of the commercial hardwood forests have emerged as internationally valuable reserves (such as the Rio Bravo area

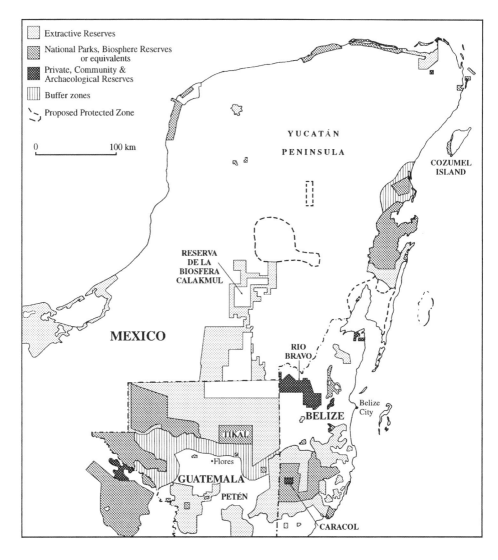

FIGURE 8. National Parks, Reserves and Protected Areas in the Maya lowlands (after Evaluacion de la Conservacion en la Selva Maya 1966).

(Burley (1992) of the Programme for Belize). At the same time and over the past twenty years, there has been a national effort at conservation and more sustainable use of forests which has been greatly aided by international agencies. There is an evident consciousness about the value of the environment and ecology as a resource which has been encapsulated in the development of eco-tourism as a major plank in government strategy.

The principal determinants of the forest cover are climate, elevation and soil type. The marked seasonality typical of the northern Yucatan peninsula affects the northern part of Belize (*c*. 1500mm, with dry months from January to May). This seasonality diminishes to some extent with altitude (as in the Maya Mountains) but more conspicuously towards the south, where rainfall is three times that of the border region with Mexico (*c*. 4500mm and the dry season is much less evident (Walker 1973). The Maya Mountains reach heights of over 1200m and this is sufficient to maintain the cooler temperatures typical of the southern extensions of the nearctic temperate forest (characterised by oaks, notably *Quercus oleiodes*). The extensive montane pine savannas are at lower altitudes than their Mexican counterparts and reflect soil conditions as much as elevation (Furley 1974, 1976). The extremely old land surface and highly weathered metamorphic and granitic Oxisols and Ultisols (Baillie *et al*. 1993) provide little in the way of plant nutrients, and are notably deficient in phosphorus and other organic elements whilst maintaining high levels of aluminium. At times even the pines are absent leaving bare *Andropogon-Eragrostis* grasslands (the Bald Hills). These may be partly a reflection of periodic intense storms, such as the devastating legacy of Hurricane Hattie in 1962, and of fires. Elsewhere around the Maya Mountains, where soil parent materials are more favourable (as on limestones or shales, Furley and Newey 1979), semi-evergreen forests predominate, passing into more permanent evergreen forest towards the southern border. Over the lowlands, a similar semi-evergreen forest is typical of moist sites on richer soils, becoming increasingly deciduous northwards (Pennington and Sarukhan 1968; Iremonger and Brocaw 1994), whilst on the outwash sands and gravels from the Maya Mountains and on the old coastal terraces, now many miles inland, a lowland form of pine savanna is repeated in well drained areas or palmetto palm savannas in wetter tracts. Finally along the coasts and in numerous inland lagoons and wet depressions, wetlands prevail with widespread mangrove communities which extend out onto the barrier reef cayes and further atolls (Furley and Ratter 1992; Zisman 1992).

Yet the apparent longevity of these diverse formations is an illusion. The vegetation cover has been extremely dynamic, reflecting both environmental and human change. Whilst the Maya Mountains may have remained relatively stable over immense periods of time, the lowlands have experienced successive phases of emergence from the sea, with the remains of old coastlines, rivers and lagoons. The coastal zone processes been highly active over the Holocene, but terrestrial processes have also been important. Quantities of coarse textured sands and gravels have been eroded from the mountains and then deposited and re-sorted by vigorous radial stream systems over much of the Belize River valley and across the Southern lowlands. Equally, the impact of the Maya can hardly be underestimated and evidences of the occupation are constantly being discovered throughout the length and breadth of the country, even in areas which appear remote and unpromising today. This is eloquently reflected in the title of the book by Turner (1983) on the Maya way of life: 'Once beneath the forest'. Although palaeoenvironmental evidence is weak for Belize, scattered information from the Mayan area as a whole supports the view that the Pleistocene dryness and slightly cooler temperatures had severely reduced tropical features in the forest and that the present day conditions gradually evolved during the Holocene. In the Lake Peten area of northern Guatemala for example, the Late Glacial vegetation seems to have consisted of marsh, savanna and juniper scrub, which gave way to temperate forest in the early Holocene which in turn preceded the mesic tropical forest characterised by *Brosimum alicastrum* (the ramon) which we can identify today (Leyden 1984). The so-called 'primeval rain forests'so characteristic of the lowland Maya region are therefore, no older than *c*.11,000 years and many are much younger as a result of Mayan clearance and disturbance (Vaughan *et al.* 1985).

7.   Issues of contemporary forest management in Belize

This brings us to the question of contemporary forest management and the challenges and opportunities that it faces. The main concerns within Belize are the natural demands for subsistence and employment which go alongside strong support for the view that sustaining the natural resources is the best means of achieving this end. This view is supported particularly in Mexico, Guatemala and Belize where 'La Ruta Maya' has been initiated specifically to provide a 'routeway' for tourists looking to combine archaeological and envionmental interests.

        The Mayan forest areas in these countries have been recently evaluated for conservation purposes at a regional workshop held in Chiapas. Groups from the three countries, plus observers and participants from international agencies, combined to give an overview of the conservation status and problems of the Maya Forest (Evaluacion de la Conservacion en la Selva Maya 1995). Nineteen priority areas have been identified for the region based on the level of threat, the distribution and biodiversity, the critical nature of the habitats, the level of endemism and the overall ecological importance (Figure 9a). Eight of these lie completely within the territory of Belize or shared with immediate neighbours and five are considered to be of very high priority. The largest is shared with Guatemala and Mexico (Tikal-southern Calakmul-Rio Bravo and extending across the northern Peten). Other areas of high priority in Belize include Crooked Tree Lagoon, the Northern and Southern Lagoons and locations in the southern Maya Mountains (shared with Guatemala). A number of vitally important biological reserves were also identified, some being migration routes, some riparian zones or wetlands, others buffer zones or competitive management zones. These are all regarded as critical and highly sensitive areas demanding particular care in land management (Figure 9b). Towards this end, a further initiative has been the establishment of a database for the Mayan region, a joint project between the US Man and the Biosphere Programme, ProPeten and Conservation International.

        Much of this area (Figure 10) is already designated for natural parks or equivalent reserves (terrestrial and marine) or as extractive reserves or private reserves. Although numerous archaeological sites have been discovered and individual locations are protected privately or nationally, there is only one 'anthropological reserve' so far designated - the enormous and mainly unrestored complex at Caracol in the Maya Mountains. Whilst the total picture on paper at least, seems relatively healthy (Zisman 1996), there are a number of growing concerns over the effectiveness of some of the reserves and their permanence in the face of increasing pressures on the land (Belize Tropical Forest Action Plan, ODA 1989). At the moment, from a conservationist

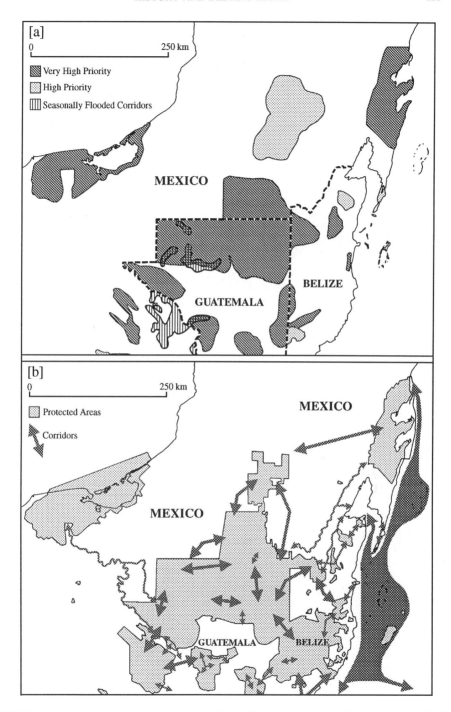

FIGURE 9. Conservation of the remaining Maya forest (a) biological resources, and (b) biological corridors (after Evaluacion de la Conservacion en la Selva Maya 1966).

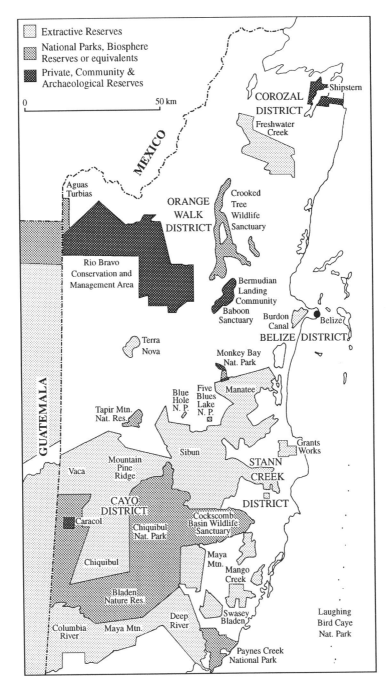

FIGURE 10.   The conservation status of natural vegetation in Belize (modified from Evaluacion de la Conservacion en la Selva Maya 1966).

point of view, a reasonably favourable balance has been struck between the needs of smallholder farmers (particularly the one fifth to one sixth who may be refugees - the numbers of which are not at all certain) and the areas delimited as reserves. For example, whereas much of the littoral forest (transitional broadleafed trees on calcareous sands of low terraces) is under great threat, much of the mangrove forest along the coast is still intact (c. 90-95%) except for the rapidly expanding area around Belize City. However, many of the more popular cayes and vulnerable coastal locations are in private hands with little government control over conservation. Furthermore it has proved difficult to assess the real economic value of the wetland component of the coastal zone (McField et al. 1996).

One major problem confronting the promotion of sustainable forestry is the minor role it plays in the national GDP (around 2%). This figure does not, however, include sawmilling or complex values such as its contribution to the ecotourist industry. Unfortunately forests have been so widespread that they have been perceived as an endless resource and their fate has not received the same attention that it has gained elsewhere (Hecht and Cockburn 1989). Furthermore successive reports appear to have exaggerated the forest resources and productivity leading to what has been argued as a misleading national tropical forest action plan (Howell 1994). For instance Howell cites official figures for the export of mahogany which are more than double the figures recorded as logged. This has led to a lack of drive or feeling of urgency in replenishing forest stocks or by seeking alternative complementary strategies such as agroforestry. According to Howell, there simply is no surplus forest land available for agriculture and he signals the urgency of national protection measures for both state and private forests. Forest policy still follows the colonial guidelines of 1954 which contained a number of ideals to which few could object but contained little consideration of practical strategies. To some extent this is being addressed in the current Forest Planning and Management Project (Belize Forest Department 1993), where systematic monitoring has been started on a series of permanent one hectare plots on state land and far more attention is given to long term planning (see for example Bird 1994).

8. Forest management in the Selva Maya: issues and alternative strategies

The issues facing the management of the forests within the Maya lowlands are consonant with those of many parts of the humid tropics yet different in respect of a different land use history, culture and potential.

These issues may be grouped into those dealing with existing forest such as sustainability and conservation, and those treating with the causes and effects of deforestation and disturbance. Underpinning all such issues are changes over time and space and contemporary human dimensions including poverty, population increases, landlessness or land insecurity. As has been frequently demonstrated, there is an inverse relationship between population and forest cover but this simple association camouflages a multitude of underlying reasons for forest depletion. The arguments have been addressed from ends of a spectrum of stakeholders' interests. In 'Understanding Central America' (Booth and Walker 1993), for instance, there is practically no mention of forest nor much attention paid the the possible role of the environment; equally there are numerous examples of forest policies which hardly consider people. The present argument is that all approaches within this spectrum need to be accommodated: 'unless we address the growing and legitimate needs of rural people, there is little chance of turning round the rampant deforestation in Central America' (Hartshorn 1989).

The question of *sustained use* of land has received prodigious attention without generating a concensus of opinion on the policies to be adopted. A number of preferred strategies have nevertheless been identified and some combination of these is likely to provide alternatives for different countries. As Dalfelt (1991) has remarked 'the issue of sustainable use has to be defined in relation to context, whether wood production and supply, ecosystem conservation or economically'. To this could be added the specific interests of local people and of social justice.

The principle of maximum sustainable yield has been a long held ideal in forestry but foresters, conservationists, economists, politicians and local people all approach the issue from a different perspective. If sustainability of the ecosystem is the main objective, then maintentance of species diversity and its associated mutualistic interactions plus the processes which support the systems, need emphasis. This involves an understanding of regeneration, succession and homeostasis, implicitly including dynamic change, without reducing the conditions for restoration. Examples include the preservation of seedbanks, the maintenace of forest patches, mosaics and biological corridors (Schelhas 1994; Schelhas and Greenberg 1995). If sustainable timber production is paramount, then the issues revolve around markets and silvicultural methods and the inescapable fact that natural mixed forest has a considerably lower productivity than commercial productivity (Dalfelt 1991) - ignoring all the energy inputs and production over a long time span. Finally if economic sustainability becomes the goal, then forest production for a limited number of viable species or extractive products needs to be set against against possible costs of changing the form of land use. Economic valuation of tropical forests is still finding its way but each successive method seems to accommodate a greater number of considerations into the balance sheet and may forseeably approach the valuation given by conservationists: 'the economic interest of preservation...[is] ..not yet firmly established '(FAO 1993). At the same time, attempts are being made to restore degraded land and regenerate natural forest. One of the best known and longest running of such projects is that of the Guanacaste dry forest in Costa Rica where 700 km^2 is under restoration (Janzen 1986). What becomes evident from the literature is that some form of multiple use is the only avenue likely even partly to satisfy such divergent perceptions of forest resources.

Lip service is frequently paid to the ideals of conservation without the political will or economic muscle to put them into effect. There is a pressing need to classify and quantify and possibly prioritise the biological resources within the environments that sustain them.   Most commentators now accept that successful biological conservation necessitates the involvement and incorporation of local people. In Central America this may include such disparate groups as indigenous Indians, mestizos, creoles, garifunas and smallholder colonists as well as larger scale landowners. Traditional knowledge has become much better recognised and valued (Posey 1985, 1992; Gomez-Pompa and Kaus 1990; Schmink *et al.* 1992) raising the issues of intellectual property rights and stewardship of forests.

The link between resource preservation and resource use is well illustrated in several of the Central American protected areas. In the Guatemalan Peten for example, the Reserva de la Biosfera Maya incorporates five new national parks and a single multipurpose reserve (Nations 1992) and with an economy mostly based on extractive products, notably the xate palm, chicle and allspice along with growing eco-archaeological tourism (see Figure 8). Analysis has shown that comparing cattle ranching and agricultural colonisation with resource extraction, over a 12 year planning forecast and a real discount rate of 5%, the first two land use options had negative net present value and cost benefit ratio and were therefore non-sustainable (Nations 1992). To ensure conservation, it will be necessary to demonstrate that the economic value, however derived, of undisturbed forest is greater than that of any replacement land use amortised over the longer term, whilst at the same time restraining demand for forest products beyond their reproductive capability. A potential source of added value is the increasing realisation of ethno-botanical products; for instance Toledo *et al.* (1992) report on nearly 2000 products from just over 100 useful plant species in Mexican tropical forests. Lugo and Lowe (1995) refer to a 'worst case scenario' where tropical forests are fragile, population growth produces demands beyond the capacity of the forests, along with a mixture of human greed, misguided public policies and the failure of the market system to act as a control. He goes on to argue that this doomwatch prediction can be considerably mollified by better organisation of land and concentration of human activities, with greater understanding of the true resilience of forests, improved management practices, education and training. Effectively this comes down to more knowledeable utilisation of the forest resources.

Literature on forest clearance and disturbance has reached such a saturation point that it is less a question of information or strategies but one of implementation of policies known to be effective. The causes are nevertheless worth re-visiting briefly and can be simplified into direct and underlying. The direct causes are expansion of agriculture including grazing of livestock, excessive extraction of fuelwood, commercial logging construction and infrastructural development. In Latin America, government-inspired, private or spontaneous smallholder agriculture has been reported to be responsible for 35% of tropical deforestation (Rowe *et al.* 1992); although this has been queried in some regions where larger scale landowners are more actively responsible and, locally, logging or other commercial interests may be dominant (for example in Costa Rica, see Lutz and Daly (1991)). In any case the figures represent landlessness and lack of land security and are therefore derived from deeper social causes. The underlying causes that are frequently cited are rural poverty, pressures produced by population growth, misguided public policies and the weak state of most of the national economies in the region. Inherent failures in the world free market affecting forest use and management are related to 'externality costs', the problems of economic valuation of forests and the conflicts and time scale between present and future needs together with a lack of clear definition of property rights (Rowe *et al.* 1992). There are therefore urgent economic, social as well as environmental concerns.

Since tropical forest management has to fit real needs within each social and cultural context (Gomez-Pompa and Bainbridge 1995), then different forests or zones within forests are likely to require different skills and arguably policy changes (Current and Scherr 1995). Within the Mayan forests it might be possible, following the policy suggested by the FAO (1993), to shape management strategies to population density. In areas of high density and scarce land resources, there are likely to be fuelwood and timber shortages requiring protection forests, intensive management and agroforestry techniques. At lower population densities with greater land reserves, zoning could include logging, using more sustainable practices, with some extractive products utilising ethnobotanical knowledge, along with the conservation of critical areas. This is the situation in much of the peripheral Maya forest area, in the southern Peten, western Honduras, northern Mexican Yucatan and coastal/riverine Belize. Finally in the core areas of forest with sparse population and the bulk of the remaining reserves, there is scope for greater level of conservation and time to develop technical management techniques with an opportunity to develop extractive products. This situation pertains in the Biosphere Reserves and protected areas of the Mayan forest heartland (US MAB/ProPeten/Conservation International 1995), see Figure 8.

Of the alternatives that are present in these three contexts, protection forests, plantations and intensive agroforestry require high levels of management skill and high level of cooperation and involvement with local people, drawing on indigenous knowledge of the plants and environment. There is likely to be little scope for conservation in such pressurised forest areas, although education and training and making use of the natural skills and traditions of local people could contribute greatly. The focus might be described as household and community ecology, with kitchen gardens, living field boundaries, use of shade trees and coppicing (Bullock *et al.* 1995). With less pressure, plantations affording commercial return could be interpolated into a multiple forest use, possibly with emphasis on successional stages of forest regrowth or silvicultural schemes such as *taungya* which has similar principles, together with various forms of agroforestry. At the lowest population pressures, there is perhaps the only opportunity remaining for testing out new or revived ideas on balanced forest strategies, whilst retaining a conservation core. At each of these levels of forest activity, four components have been repeatedly identified: firstly physical and biological, secondly social (including political and cultural elements), thirdly economic aspects (including costs and prices, rates of return, markets, trade), and finally, technology (comprising forestry practices as well as harvesting and processing) Research needs to be continued on all of these fronts. At present, the resources given to national forest agencies and institutions are generally inadequate, either for the research required or for the execution and monitoring of forest policy (Lugo and Lowe 1995). Because of this, international and regional agencies have been extremely significant in developing research and stimulating a change of direction. The role for example, of CATIE (Centro Agronomico Tropical de Investigacion e Ensenanza) or of OTS (Organisation of Tropical Studies), with their biological station at La Selva (north east Costa Rica, McDade *et al.* 1994), have been outstanding.

In addition to its biological importance, the core areas also contain numerous spectacular Mayan archaeological sites, such as Tikal in Guatemala, Caracol in Belize and Calakmul in Mexico. The combination of biological richness and historic interest provides the basis for an unrivalled eco-tourist industry which, if carefully controlled, could generate the economic value required to counterbalance replacement land uses.

9.   Conclusions

The *Selva Maya* remains one of the most important tropical forest reserves in Middle America. The core areas are biological rich and coincide with the location of major archaeological sites. This may provide an income from eco-archaeological tourism which can match or exceed the short term gains possible from forest clearance. It might provide breathing space for the development of a multipurpose range of alternative strategies for forest management.

At present information is marginal for the development of such programmes. The level of knowledge on plant or animal taxonomy, ethnoecology, biotic changes and dynamic processes between organisms and environment is still imprecise, as is an understanding of long term change. Additionally, the degree of involvement of local people and understanding of their social and cultural traditions is still meagre. This demands attention on several fronts: the establishment of long term and well monitored experiments, better understanding of palaeoecological changes during the Holocene and more vigorous attempts to include and utilise local knowledge incorporating local people. To tackle the threat to the long term survival of these forests not only requires fresh ideas but also a reduction in the underlying causes of depletion. Above all it requires political will at all levels from national governments to community leaders.

Note

1. Assuming that the conditions resembled those of today, the edaphic characteristics would have been highly variable, ranging from the typical dark organic mollisols developed on limestones, often with vertic or swelling and shrinking clays, to the sandy, porous and low nutrient inceptisols of outwash deposits, and from excessively to very poorly drained profiles.

References

Abrams, E.M and Rue, D.J. (1988). The causes and consequences of deforestation among the prehistoric Maya, *Human Ecology*, **16** (4), 377-395.

Baillie, I.C.,Wright., A.C.S., Holder, M.A. and Fitzpatrick, E.A. (1993). Revised classification of the soils of Belize, *Bulletin of the Natural Resources Institute*, ODA, 59, Chatham,UK..

Barrera, A.M., Gomez-Pompa, A. and Vazquez, C. (1977). El manejo de las selvas por los Mayas: sus implicaciones silvicolas agricolas, *Biotica*, **2**, 47-60.

Bateson, J.H. and Hall, I.H.S. (1977). *The Geology of the Maya Mountains, Belize*, Institute of Geological Sciences, London, Overseas Memoir, 3, 1-43.

Beard, J.S. (1944). Climax vegetation in tropical America, *Ecology*, **25**, 127-58.

Beard, J.S. (1955). The classification of tropical American vegetation types, *Ecology*, **36**, 89-100.

Belize Forest Department (1993). *The Forests of Belize: a first approximation at estimating the country's forest resources*, Forest Department, Ministry of Natural Resources. Belmopan, Belize (ms 10pp).

Beltran, E. (ed.) (1959). *Los Recursos Naturales del Sureste y su Aprovechamiento*, Ediciones INMERNAR, Mexico City.

Bird, N.M. (1994). *Draft Forest Managemen tPlan: Colombia Forest Management Unit*, Ministry of Natural Resources, Belmopan, Belize.

Booth, J.A. and Walker, T.W. (1993). *Understanding Central America*, 2 ed.,Westview Press, Boulder, Colorado.

Bullock, S.H., Mooney, H.A. and Medina, E.(eds.) (1995). *Seasonally dry forests*, Cambridge University Press, Cambridge.

Burley, F.W. (1992). The Rio Bravo Conservation and Management area in Belize, in K.H. Redford and C. Padoch (eds.) *Conservation of Neotropical Forests*, Columbia University Press, Cambridge, pp. 1220-1227.

Bush, M.B., Piperno, D.R., Colinvaux, P.A., de Oliviera, P.E., Krissek, L.A., Miller, M.C. and Rowe, W.E.

(1992). A 14,300 year palaeoecological record of a lowland tropical lake in Panama, *Ecological Monographs*, **62**, 251-275.

Chapman,V. (1977). *Wet Coastal Ecosystems*, Elsevier, Amsterdam.

Coney, P.J. (1982). Plate tectonic constraints on biogeographic connections between North and South America, *Annals of Missouri Botanic Garden*, **69**, 432-443.

Current, D. and Scherr, S.J. (1995). Farmer costs and benefits from agroforestry and farm forestry projects in Central America and the Caribbean: implications for policy, *Agroforestry Systems*, **30**, 87-103.

Dalfelt, A. (1991). Ecological constraints to sustainable management of the tropical moist forest, Divisional Working Paper No. 1991-25, Policy & Research Division, Environment Department, World Bank, Washington

Darch, J.P. (ed.) (1983). *Drained Field Agriculture in Central and South America*, British Archaeological Reports International Series No. 189, B. A. R., Oxford.

D'Arcy, W.G. (1977). Endangered landscapes in Panama and Central America; the threat to plant species, in G.T. Prance and T.S. Elias (eds.) *Extinction is Forever*, New York Botanical Garden, pp. 89-102.

Davis, S.D., Heywood, V.H. and Herrera-Macbryde, O. (eds.) in press. *Centres of Plant Diversity: a guide and strategy for their conservation, vol. 3. The Americas*, World Wide Fund for Nature and IUCN/The World Conservation Union.

Evaluacion de la Conservacion de la Selva Maya. (1995). Map: scale 1:800,000 and commentary, Conservation International, U.S. Man and the Biosphere Program & others, Washington.

F.A.O. (1993). *Forest Resources Assessment 1990: tropical countries*, Forestry Paper 112, F.A.O., Rome.

Flenley, J.R. (1979). *The Equatorial Rain Forest: a geological history*, Butterworths, London.

Flenley, J.R. (1992). Palynological evidence relating to disturbance and other ecological phenomena of rain forests, in J.G. Goldhammer (ed.), *Tropical forests in transition*, Birkhauser, Basel, pp. 17-24.

Flores, S. and Ucan, E.E. (1983). Nombres usados por los Mayas para designar a la vegetacion, Cuadernos de Divulgacion, *INIREB*, **10**, 1-33, Mexico City.

Furley, P.A. (1974). Soil-slope-plant relationships in the northern Maya Mountains, Belize, Central America. I. The sequence over metamorphic rocks and sandstones, *Journal of Biogeography*, **3**, 171-186.

Furley, P.A. (1974). Soil-slope-plant relationships in the northern Maya Mountains, Belize, Central America. II.. The sequence over phyllites and granites, *Journal of Biogeography*, **3**, 263-279..

Furley, P.A. (1976). Soil-slope-plant relationships in the northern Maya Mountains, Belize, Central America. III. Variations in the properties of soil profiles, *Journal of Biogeography*, **3**, 303-319.

Furley, P.A. and Newey,W.W. (1979). Variations in plant communities with topography over tropical limestone soils, *Journal of Biogeography*, **6**, 1-15.

Furley, P.A., Munro, D.M., Darch, J.P. and Randall, R.R. (1995). Human impact on the wetlands of Belize, Central America, in R.A Butlin and N. Roberts (eds.) *Ecological Relations in Historical Time*, Blackwell, Oxford, pp. 280-307.

Furley, P.A. and Ratter, J.A. (1992). *Mangrove Distribution, Vulnerability and Management in Central America*, ODA-OFI Forestry Research Programme R.4736, Edinburgh.

Gentry, A.H. (1992). Tropical forest biodiversity; distributional patterns and their conservational importance, *Oikos*, **63**, 19-28.

Gentry, A.H. (1995). Diversity and floristic composition of neotropical dry forests, in S.H. Bullock, H.A. Mooney and E. Medina (eds.) *Seasonally dry forests*, Cambridge University Press, Cambridge, pp. 146-194.

Gliessman, S.R. (1990). *Agroecology. Researching the Basis for Sustainable Development*, Springer Verlag, New York.

Goldhammer, J.G. (ed.) (1992). *Tropical Forests in Transition*, Birkhauser, Basel.

Gomez-Pompa, A. (1987). On Maya silviculture, *Mexican Studies*, **3** (1), 1-17.

Gomez-Pompa, A. (1991). Learning from traditional ecological knowledge: insights from Mayan silviculture, in A. Gomez-Pompa, T.C. Whitmore and M. Hadley, (eds.) *Rain Forest Regeneration and Management*, UNESCO-Parthenon Press, pp. 335-342.

Gomez-Pompa, A. and Kaus, A. (1990). Traditional management of tropical forests in Mexico, in A.B. Anderson (ed.) *Alternatives to Deforestation; steps towards sustainable use of the Amazon rain forest*, Columbia University Press, New York, pp. 43-64.

Gomez-Pompa, A. and Bainbridge D.A. (1995). Tropical forestry as if people mattered, in A.E. Lugo and C. Lowe (eds.) *Tropical Forests: management and ecology*, Springer Verlag, New York, pp. 408-422.

Gomez-Pompa, A., Flores, E. and Sosa, V. (1987). The 'pet kot': a man made tropical forest of the Maya, *Interciencia*, **12** (1), 10-15.

Graham, A. (1992). Utilization of the isthmian land bridge during the Cenozoic - palaeobotanical evidence for timing and the selective influences of altitude and climate, *Review of Palaeobotany and Palynology*, **72**, 119-128.

Graham, A. and Dilcher, D. (1995). The Cenozoic record of tropical dry forest in northern Latin America and southern United States, in S.H. Bullock, H.A. Mooney and E. Medina (eds.) *Seasonally Dry Tropical Forests*, Cambridge University Press, Cambridge, pp. 124-145.

Hammond, N. (1978). The myth of the milpa: agricultural expansion in the Maya lowlands, in P.D. Harrison and B.L. Turner (eds.) *Pre-Hispanic Maya Agriculture*, University of New Mexico Press, Alburqueque, pp. 23-34.

Harcourt, C.S. and Sayer, J.A. (1996). *The Conservation Atlas of Tropical Forests: The Americas*, Simon and Schuster, New York.

Harrison, P.D. (1978). So the seeds shall grow: some introductory comments, in P.D. Harrison and B.L. Turner (eds.) *Pre-Hispanic Maya Agriculture*, University of New Mexico Press, Alburqueque, pp. 1-11.

Harrison, P.D and Turner, B.L. (eds.) (1978). *Pre-hispanic Maya Agriculture*, University of New Mexico Press, Albuqueque.

Hartig, H.M. (1979). *Las aves de Yucatan*, Fondo Editorial de Yucatan, Cuaderno 4. Merida, Mexico.

Hartshorn, G.S. (1989). Forest loss and future options in Central America, in *Manomet Symposium on the Ecology and Conservation of neotropical migrant landbirds*, Manomet, Boston, pp. 13-19.

Hecht, S. and Cockburn, A. (1989). *The Fate of the Forest*, Penguin Books, London.

Holdridge, L.R. (1967). *Life Zone Ecology*, Tropical Science Center, San Jose, Costa Rica.

Holdridge, L.R. (1971). *Forest Environments in Tropical Life Zones: a pilot study*, Pergamon Press, Oxford.

Howell, J.H. (1994). *Belize Tropical Forest Plan*. First Quinquennial review Document 1989-94, Forest Management Division, Ministry of Natural Resources, Belmopan, Belize.

Iremonger, S. and Brokaw. N. (1994). *Vegetation Classification for Belize*, Report to NARMAP, Belmopan, Belize.

Janzen, D.H. (1986). *Guanacaste National Park: tropical ecological and cultural restoration*, Editorial Universidad Estatal a Distancia, San Jose, Costa Rica.

Janzen, D.H. (1988). Tropical dry forests:the most endangered major tropical ecosystem, in E.O. Wilson (ed.) *Biodiversity*, National Academy Press, Washington, pp. 130-137.

Kellman, M., Tackaberry, R. and Meave, J. (1996). The consequences of prolonged fragmentation; lessons from tropical gallery forests, in J. Schelhas and R. Greenberg (eds.) *Forest Patches in Tropical Landscapes*, Island Press, Washington, pp. 37-58.

Lauer,W. (1968). Problemas de la division fitogeografica en America Central, in C. Troll (ed): *Geoecology of the Mountainous Regions of the Tropical Americas,* Proceedings of the UNESCO-Mexico Symposium 1966, Colloquium Geographicum Band 9, 139-56.

Leonard, H.J. (1987). *Natural Resources and Economic Development in Central America: a regional environmental profile*, Transaction Books, New Brunswick.

Leyden, B.W. (1984). Guatemalan forest synthesis after Pleistocene aridity, *Proceedings National Academy Sciences*, USA, **91**, 4856-4859.

Lugo, A.E. (1990). Fringe wetlands, in A.E. Lugo, M. Brinson and S. Brown (eds.) *Forested Wetlands*, Elsevier, Amsterdam, pp. 143-170.

Lugo, A.E., Brinson, M. and Brown, S. (1990). *Forested Wetlands*, Elsevier, Amsterdam.

Lugo, A.E., Schmidt, R. and Brown, S. (1981). Tropical forests in the Caribbean, *Ambio,* 10 (6), 319-324.

Lugo, A.E. and Lowe, C. (eds.) (1995). *Tropical Forests: management and ecology*, Springer Verlag, New York.

Lundell, C.L. (1940). The 1936 Michegan-Carnegie Botanical Expedition to British Honduras, in Botany of the Maya area, Miscellaneous Papers 14, Carnegie Institute, Washington, Publication No. 522.

Lutz, E. and Daly, H. (1991). Incentives, regulations and sustainable land use in Costa Rica, *Environmental and Resource Economics*, **1**, 179-194.

McDade, L.A., Bawa, K.S., Hespenheide, H.A. and Hartshorn, G.S. (eds.) (1994). *La Selva: ecology and natural history of a neotropical rain forest*, University of Chicago Press, Chicago.

MacNeish, R.S. (1992). *The Origins of Agriculture and Settled Life*, University of Oklahoma Press, London.

McField, M., Wells, S. and Gibson J. (eds.) (1996). State of the coastal zone report, Belize 1995, UNDP, Belize City (Project Bze/92/G31).

Meave, J. and Kellman, M. (1994). Maintenance of rain forest diversity in riparian forests of tropical savannas; implications for species conservation during Pleistocene drought, *Journal of Biogeography*, **21**, 121-135.

Mooney, H.A., Bullock, S.H. and Medina, E. (1995). Introduction, in S.H. Bullock, H.A. Mooney and E. Medina (eds.) *Seasonally Dry Tropical Forests*, Cambridge University Press, Cambridge, pp. 1-8.

Murov, N.T. (1967). *The Genus Pinus*, Ronald Press, New York.

Murphy, P.G. and Lugo, A.E. (1995). Dry forests of central America and the Caribbean, in S.H. Bullock, H.A. Mooney and E. Medina (eds.) *Seasonally Dry Tropical Forests*, Cambridge University Press, Cambridge, pp. 9-34.

Nations, J.D. (1992). Xateros, chicleros and pimenteros: harvesting renewable tropical forest resources in the Guatemalan Peten, in K.H. Redford and C. Padoch (eds.) *Conservation of Neotropical Forests*, Columbia University Press, New York, pp. 208-219.

Nations, J.D. and Komer, D.I. (1983). Central America's tropical rain forests; positive steps for survival, *Ambio*, **12** (5), 233-239.

ODA (Overseas Development Administration). (1989). Belize Tropical Forest Action Plan, ODA, London.

Pennington, T.D. and Sarukhan, J. (1968). *Arboles Tropicales de Mexico*, INAF/FAO. Mexico.

Perry, J.P. (1991). *The Pines of Mexico and Central America*, Timber Press, Oregon.

Plotkin, M. and Famolare, L. (eds.) (1992). *Sustainable Harvest and Marketing of Rain Forest Products*, Island Press, California.

Pohl, M. (ed.) (1985). *Prehistoric Lowland Maya Environment and Subsistence Economy*, Peabody Museum, Harvard University Press.

Posey, D.A. (1985). Indigenous management of tropical forest ecosystems. The case of the Kayapo Indians of the Brazilian Amazon, *Agroforestry Systems*, **3** (2), 139-158.

Posey, D.A. (1992). Traditional knowledge, conservation and the rain forest harvest, in M. Plotkin and L. Famolare (eds.) *Sustainable Harvest and Marketing of Rain Forest Products*, Island Press, California. pp. 46-50.

Redford, K.H. and Padoch, C. (eds.) (1992). *Conservation of Neotropical Forests*, Colombia University Press, New York.

Ratter, J.A. (1992). Transitions between cerrado and forest vegetation in Brazil, in P.A. Furley, J. Proctor and J.A. Ratter (eds.) *Nature and Dynamics of Forest-Savanna Boundaries*, Chapman & Hall, London, pp. 417-430.

Rowe, R., Sharma, M. and Browder, J. (1992). Deforestation: problems, causes and concerns, in N.P. Sharma (ed.) *Managing the World's Forests*, Kendell-Hunt Publishing Company, Dubuque, Iowa, pp. 33-46.

Rudel, T. and Roper, J. (1996). Regional patterns and historical trends in tropical deforestation 1976-90: a qualitative comparative analysis, *Ambio*, **25** (3), 160-166.

Saenger, P., Heger, I.E.. and Davie, J. (eds.) (1983). The global status of mangrove ecosystems, *The Environmentalist* (3), Supplement No.3.

Schelhas, J. (1994). Building sustainable land uses on existing practices; smallholder and use mosaics in tropical lowland Coast Rica, *Society and Natural Resources*, **7**, 67-84.

Schelhas, J. and Greenberg, R. (eds.) (1995). *Forest Patches in Tropical Landscapes*, Island Press, Washington.

Schmink, M., Redford, K.H. and Padoch, C. (1992). Traditional peoples and the biosphere; framing the issues and defining the terms, in K.H. Redford and C. Padoch (eds.) *Conservation of Neotropical Forests*, Colombia University Press, New York, pp. 3-13.

Siemens, A.H. and Puleston, D.E. (1972). Ridged fields and associated features in southern Campeche. New perspectives on the lowland Maya, *American Antiquity*, **37**, 228-239.

Sosa,V., Flores, J.S., Rico-Gray, V., Lira, R. and Ortiz, J.T. (1985a). *Lista floristica y sinonimia Maya*, Etnoflora Yucatanense, Fasc.1, INIREB, Mexico.

Sosa,V., Gomez-Pompa, A and Flores, S. (1985b). La flora de Yucatan, *Ciencia y Desarrollo*, **60**, 37-46.

Stadtmuller, T. (1987). *Cloud Forests of the Humid Tropics*, U.N. University and CATIE (Centro Agronomico Tropical de Investigacion e Ensenanza), Turrialba, Costa Rica.

Toledo,V.M. (1982). Pleistocene changes of vegetation in tropical Mexico, in G.T. Prance (ed.) *Biological Diversification in the Tropics*, Colombia University Press, New York, pp. 93-111.

Toledo,V.M., Batis, A.I., Becerra, R., Martinez, E. and Ramos, C.H. (1992). Products from the tropical rain forests of Mexico; an ethnoecological approach, in M. Plotkin and L. Famolare (eds.) *Sustainable Harvest and Marketing of Rain Forests Products*, Island Press, California, pp. 99-109.

Tomlinson, P. (1986). *The Botany of Mangroves*, Cambridge University Press, Cambridge.

Turner, B.L. (1983). *Once Beneath the Forest*, Bowker Publishing Co., London.

Turner, B.L and Denevan W.M.(1989). Prehistoric manipulation of the wetlands in the Americas: a raised field perspective, in I. Farrington (ed.) *Prehistoric Intensive Agriculture in the Tropics*, British Archaeological Reports International Series, B.A.R., Oxford, pp. 16-48.

Turner, B.L and Harrison, P.D.(eds.) (1983). *Pulltrouser Swamp: ancient Maya habitat, agriculture and settlement in northern Belize*, University of Texas Press, Austin.

US MAB/ProPeten/Conservation International. (1995). *Maya Tri-National Forest Meta Data Survey*, Regional Conservation Analysis Program, Conservation International, Washington.

Vaughan, H.H., Deevey, E.S. and Garrett-Jones, S.E. (1985). Pollen stratigraphy of two cores from the Peten lake district, with an appendix on two deep-water cores, in M. Pohl (ed.). *Prehistoric Lowland Environment and Subsistence Economy.* Papers of the Peabody Museum of Archaeology and Ethnology, Harvard University, Cambridge, Massachussetts, **77**, 73-89.

Vuilleumier, F. and Monasterio, M. (eds.) (1986). *High Altitude Tropical Biogeography*, Oxford University Press.

Walker, S.H. (1973). *Summary of Climatic Records for Belize*, Supplementary Report No.3., Land Resources Division, Ministry of Overseas Development, London.

West, R.C. and Augelli, J.P. (1986). *Middle America*, Prentice-Hall, Eaglewood Cliffs, New Jersey.

Weyl, R. (1980). *Geology of Central America*, 2nd ed., Gebruder Borntraeger, Berlin.

Whigham, D.F., Zugasty Towle, P., Cabrera Cano, E., O'Neill, J. and Ley, E. (1990). The effect of variation in precipitation on growth and litter production in a tropical dry forest in the Yucatan of Mexico, *Tropical Ecology*, **31**, 23-34.

Whitmore, T .C. and Prance, G.T. (eds.) (1987). *Biogeography and Quaternary History in Tropical America*, Clarendon Press, Oxford.

Wiseman, F.M. (1978). Agricultural and historical ecology of the Maya lowlands, in P.D. Harrison and B.L.Turner (eds.) *Pre-Hispanic Maya Agriculture*, University of New Mexico Press, Alburqueque, pp. 63-116.

WRI (World Resources Institute). (1995). *World Resources 1994-95*, Oxford University Press, Oxford.

Wright, A.C.S., Romney, D.H., Arbuckle, R.H. and Vial, V.E. (1959). *Land in British Honduras*, Colonial Office Research Publication No. 24, H.M.S.O., London.

Zisman, S. (1992). *Mangroves in Belize; their characteristics, use and conservation*, Forest Planning & Management Project, Ministry of Natural Resources, Belmopan, Belize.

Zisman, S. (1996). *The Directory of Belizean Protected sites of Nature Conservation interest*, NARMAP/Government of Belize, Belmopan, Belize.

Dr. Peter A. Furley, Department of Geography, University of Edinburgh, Edinburgh EH8 9XP, Scotland.

7. SOCIAL, ECONOMIC AND POLITICAL ASPECTS OF FOREST CLEARANCE AND LAND-USE PLANNING IN INDONESIA

Alastair I. Fraser

1. Introduction

The forests of Indonesia are recognised both nationally and internationally for their biological richness. They also represent an important natural resource, which has made a major contribution to the economic development of the country over the past twenty years. Concern has been expressed in many quarters that the area of forest is being rapidly depleted, and that this many lead to the loss of potentially valuable species of plants and animals, adverse environmental consequences, and a depletion of the resource base of commercial timbers to support the forest industry.

Some economists argue that natural forests have very low yields and should therefore be liquidated to release capital which can be invested more profitably elsewhere, and land which could be used more productively. These arguments take no account of the environmental value of the forests, and may greatly underestimate the financial value of the forests if they are managed sustainably, and greater use is made of the potential harvest of non-timber forest products such as rattans, gums, resins, medicines, etc.

It has been claimed by Repetto and Gillis (1989) that the deforestation in Indonesia has been wasteful, and that the country has derived less benefit from its forests than it might have done with a less rapid but more managed rate of depletion. The author attempts to show that part of the deforestation has been a direct consequence of government policies, some relating directly to the forestry sector, and others arising from conflicting sectorial policies relating to land-use.

Indonesia is a very large and diverse country with five large islands, some 970 smaller inhabited groups of islands and around 13,670 islands in total, with a total land area of 1.93 million km^2. About 1.46 million km^2 of the island is classed as 'forest', but a proportion of this is no longer covered with trees.

2. Forest land cover during the 20th century

Historical data on the extent of forest cover is limited to records by early travellers such as Wallace (1869) in the 19th century, who travelled widely throughout the Indonesian archipelago. In 1850 he visited Java, and reported that the population was about 9.5 million persons, or about a tenth of the present population. At that time much of the island was still heavily forested, and the population was mainly restricted to the coastal plains and river valleys. The Dutch were interested in the teak forests in Java, and were logging them extensively during the 19th century (Ponder 1982). At the beginning of the 19th century a road was constructed the length of Java and rapidly extended to many then remote parts of the island. Large quantities of timber were then felled for local construction and export. By the end of the 19th century, the teak forests had been depleted to such an extent that the Dutch administrators began a plantation programme to reforest unproductive areas. Some 9000 ha of teak plantations established before the end of the 19th century still remain, and about 565,000 ha of productive teak plantations

133

B.K. Maloney (ed.), Human Activities and the Tropical Rainforest, 133-150.
© 1998 *Kluwer Academic Publishers. Printed in the Netherlands.*

which are now managed sustainably.

In addition to the managed teak plantations in Java, there are about 340,000 ha of plantations of other species, 820,000 ha of protection and conservation forest, 180,000 ha of natural forest, and about 490,000 ha of forest land not currently productive (Sutter 1989). Thus, despite the very high population density of about 800 people/km^2, the forest cover on Java still amounts to about 19% of the total land area (almost double that of the UK).

Outside Java in the main islands of Sumatra, Kalimantan (the Indonesian part of the island of Borneo), Sulawesi, Irian Jaya and the Moluccas (Maluku) the situation is very different.   There are also significant differences in the pattern and rate of development in each of the main islands where most of the tropical moist forest is concentrated.

Early travellers in Kalimantan such and Bock (1881), Beccari (1904), Lumholtz (1920) and Tellima (1938) reported very low population densities, and more or less continuous forest cover, except around settlements which were along rivers, and in the more fertile parts of South Kalimantan around Banjarmasin, Pontianak in West Kalimantan and Kutai in East Kalimantan, where population densities were higher.

The population changes will be discussed in the next section, but with a population estimated at less than 1 million persons in the whole of Kalimantan at the turn of the century this represents an average population density of about 2 people/km^2 or little more than 1 person/km^2 if the concentration around Banjarmasin is excluded.

A substantial proportion of the population of Kalimantan were nomadic hunter-gatherers until as recently as the 1950s, and cleared very little forest.   Those that practiced agriculture used mainly temporary swiddens or ladangs, and the many tribal groups had different traditions relating to land clearing and land tenure.

Taking the figures of an average household size of six people given by Colfer and Dudley (1992), and a maximum requirement of 40 ha per household to rotate ladangs mainly in secondary forest, with 1 person/km^2 or 6km^2 per household, the proportion of land area which would be cleared of primary forest would be about 6%. This figure is consistent with the estimate that in 1950 forests covered about 88% of the land area of Kalimantan or 48.2 million ha, while some 6.5 million ha was non-forest, mainly used for agriculture or tree crops.

Until the late 1960s commercial logging was mainly restricted to the accessible forest along the rivers and in the swamp forests, especially in West Kalimantan where *ramin* was plentiful. A major change took place in the late 1960s when government policy altered to allow foreign investment in commercial logging activities. As this applied throughout Indonesia it will be discussed in more detail after a brief description of the early conditions in the other main islands.

The early history of Sumatra and Sulawesi is a little different from that in Kalimantan. These islands had been, more influenced by the activities of the Dutch colonialists from an earlier date, and considerable clearance of the forests took place during the late 19th century, for the establishment of estates or plantations of crops such as cloves and spices, sugar cane, coconut, rubber and later palm oil. These in turn gave rise to some migration of people from Java to Sumatra.

At the beginning of the 20th century, the rising population in Java led to the introduction of organised migration (transmigration) by the Dutch, in particular to southern Sumatra, and later to north Sumatra. The population density of Sumatra and Sulawesi were respectively 9 and 13 persons/km^2 in 1900 and had increased to 26 and 31 respectively by 1950. During the first half of the 20th century, the population on both these islands grew more rapidly than in Java, at least in part due to migration, and this is reflected in a lower proportion of forest area than in Kalimantan by 1950, with 72% in Sumatra and 75% in Sulawesi. The counterbalancing non-forest land is equivalent to 7 ha/family in Sumatra and 5 ha/family in Sulawesi and suggests that only a moderate proportion of the population were practising shifting cultivation.

The higher level of agricultural activity and the consequent improved access and transportation led to a greater demand for timber for construction in the growing towns of Aceh, Palembang, Medan, Padang and Ujung Pandang (Makassar). The discovery of oil in Riau (eastern Indonesia) in 1872 led to localised developments which continued to expand, especially after 1920. These oil deposits were in swamp forests, and have led to the clearance of a considerable area of forest.

The soils in Sumatra are generally less fertile than those in Java, with the consequence that development of extensive irrigated rice growing has been much less prevalent, and shifting cultivation has been widely practised. In the early part of the 20th century, rubber cultivation expanded, and smallholder rubber plantations proliferated, often associated with transmigration schemes; this too resulted in clearance of the forest.

Thus many factors led to a gradual reduction of the forest area in Sumatra and Sulawesi prior to the period when large-scale commercial logging began in the late 1960s.

The two major eastern provinces in the Maluku and Irian Jaya were much less developed than the rest of Indonesia during colonial times and population densities and economic activity were low, so the forest were little disturbed before 1950, when it was estimated that they covered 85% of the Maluku and 97% of Irian Jaya.

In 1967 Act No. 5 of the Republic of Indonesia was promulgated which set out 'The Basic Forestry Law'. This law defined 'forests', and declared all forest not on privately owned land to be state forests. It also classified forests according to their function, and recognised protection forest, mainly for hydrologic purposes, production forest, nature conservation forest and recreation forest. The Act also made provision for the Minister responsible for forestry to be able to declare 'forest areas', which should be permanent forest, and for management rights on production forest areas to be given to state or private enterprises.

As a consequence of the introduction of this law, the urgent need for Indonesia to earn foreign exchange to finance development, and the rapid growth in demand for logs, especially in the expanding economies of Japan, Korea and Taiwan, there was a rush by local entrepreneurs to obtain permits for commercial logging operations. By

1970, more than 5 million ha had been awarded as concessions, and the area increased to 25 million ha in 1976 and 50 million ha by 1980. Log production mirrored this expansion, rising from an average of 2.5 million m^3 per year in the early 1960s to 10 million m^3 in 1970 and 25 million m^3 in 1980.

The rapid expansion in commercial logging activities is frequently blamed for the current high rate of deforestation in Indonesia, and this argument will be looked at in more detail in a later section. In doing so account must also be taken of the impact of the unusually large-scale forest fires in 1983, which are reported to have burnt 3.5 million ha of which 800,000 ha, 1,400,000 ha and 550,000 ha were, respectively, virgin forest, logged forest and peat swamp forest. The balance was in secondary forest and shifting cultivation areas (Lennertz and Panzer 1984).

3.   Population changes

Early records of population in Indonesia are sparse, but according to Hugo *et al.* (1987), in 1600 the population density in Indonesia was low compared with South Asia, China and Europe at that time. During the 19th century the population of Indonesia grew at a rate of about 1.1% annually, from 13 to 40 million people. In the 20th century the growth rate increased to about 1.7% and the population rose to 164 million in 1985. However, the population growth was much more heavily concentrated in Java than the rest of the country, with the result that the proportion of the total population residing in Java increased from around 40% in 1600 to about 61% in 1985.

The population growth rate in Java during the 19th century was well over 2% per annum, but has fallen below 2% since 1930. The other islands of Indonesia show the reverse, with lower growth during the 19th century, and much higher growth rates in the past 50 years. Some provinces in Sumatra and Kalimantan have experienced population growth rates of 4% since 1960. At least part of this reversal is due to migration of people from Java to the other islands.

In Java, there has been growing urbanisation during the past 50 years, with the population of Jakarta growing at around 4%, so that the latest figures show that more than 25% of Java's population lives in the urban areas. Outside Java, the progress of urbanisation has been slower, reaching only about 20% in Sumatra and Kalimantan and 15% in Sulawesi, Maluku and Irian Jaya.

The overall national growth in population is beginning to decline, and estimates of the future population vary according to assumptions regarding mortality, and fertility between 205 and 230 million by the year 2000. The estimates of the populations of each of the islands also vary according to assumptions of internal migration rates.

By the year 2000, the population density in Sumatra and Sulawesi will be similar to that in Java in the middle 19th century, while the population density in Kalimantan and Maluku will be similar to that in Sumatra and Sulawesi in 1950, and Irian Jaya will be similar to Kalimantan in 1950.

It has been shown in the section above, on the forest cover, that increasing population density is associated with decreasing forest cover, and this trend can be expected to continue.   Tables 1 and 2 below show respectively the evolution of population density in relation to forest cover for the main islands of Indonesia over the past 150 years, and the 1982 figures for population density and forest cover for the 21 provinces on the main islands.   These tables show that there is a strong negative correlation between population density and forest cover.

The results are shown graphically in Fig. 1.   The time series and the cross sectional data separately and combined. have been fitted with linear regressions, when transformed to logarithms, and the impact of the Java data has also been analysed by fitting regressions with and without it.   The data set is too limited to draw firm conclusions, but it would suggest that the elasticity of forest cover with respect to population pressure is in the range -0.15 to -0.4, or a 1% increase in population pressure would result in a decrease in forest cover of -0.15 to 0.4%.   The higher figure is obtained when Java is excluded.   In the main islands outside Java, current population growth is around 3%, with the exception of Sulawesi, and so this model would suggest that forest cover should be decreasing by about 0.9-1% annually or 900,000-1 million ha per annum if total forest cover is 100 million ha.   This is discussed further below.

TABLE 2.. Population density and forest cover percentage by Province in Indonesia in 1982, ranked in order of population density

Province	Population density (persons/km^2)	Total forest cover %
Irian Jaya	4	84
East Kalimantan	6	85
Central Kalimantan	7	73
West Kalimantan	18	59
Central Sulawesi	20	64
Maluku	22	81
Riau	25	62
Southeast Sulawesi	27	65
Jambi	29	52
Bengkulu	42	57
South Sumatra	45	33
DI Aceh	50	79
South Kalimantan	58	49
West Sumatra	84	61
North Sulawesi	85	60
South Sulawesi	101	46
North Sumatra	123	39
Lampung	145	18
East Java	632	23
West Java	680	17
Central Java	760	15

4.   Economic importance of the forests of Indonesia

Reference has already been made to the commercial importance of the teak forests of Java to the Dutch during the 19th century, and most recent discussions of forestry in Indonesia put great emphasis on the current large-scale commercial exploitation of the rainforests of the other main islands.

The economic importance of the forests is, however, not confined to commercial logging activities since the forests have and still do provide many of the basic necessities for a large proportion of the population.   The historical importance of the forests can only be surmised from considering how communities living in around

YEAR	JAVA			SUMATRA			KALIMANTAN			SULAWESI			MALUKU			IRIAN JAYA		
	Population density Persons/km2	Growth %	Forest cover %	Population density Persons/km2	Growth %	Forest cover %	Population density Persons/km2	Growth %	Forest cover %	Population density Persons/km2	Growth %	Forest cover %	Population density Persons/km2	Growth %	Forest cover %	Population density Persons/km2	Growth %	Forest Cover %
1850	72						1.8*											
1865	107	2.6																
1880							8.0*	5										
1900	219	2.1		9			2.3			13			5.9					
1930	315	1.2		17	2.1		3.9	1.7		22	1.8							
1950			8.8	26	2.1	72	6	2.2	88	31	1.7	75	7	0.3	85	1.4		97
1980	690	1.6	7.5	59	2.8	52	12	2.3	76	55	1.9	65	22	3.3	85	3.8	2.8	95
1985	755	1.8	19	69	3.1	48	14	3.1	72	60	1.7	58	30	2.1	81	6.2	3.3	84
2000 (a)	858	0.8		109	3.1		24	3.6		84	2.2							
					(a) Medan projections				1985 figure of 19% for Java includes plantations									
*	S. Kalimantan only																	

TABLE 1. Historical changes in population density and forest cover on the main islands of Indonesia.

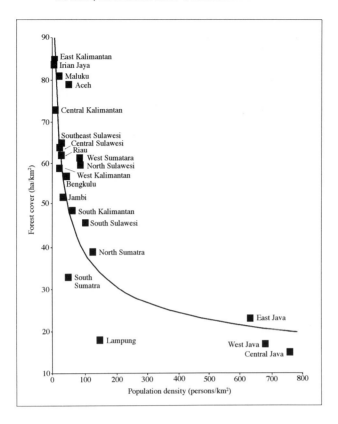

FIGURE 1. Forest cover and population density from a cross-section of Indonesian provinces.

the forest utilise the resources at the present time (cf. Ellen, this volume).

In Sumatra and Kalimantan, where most studies have been undertaken, there are many different types of communities. These have different ethnic and cultural traditions, including some which are largely forest dwelling and nomadic, such as the Kubu in Sumatra and the Penan in Kalimantan. Other groups have long practised swidden agriculture, combined with hunting, fishing and trading in forest products, while more recently migrants have settled on land cleared from forest, but have no tradition of using the forest or its products. These generalised categories represent a continuum of dependency on the forest from the almost complete reliance of the hunter-gatherers to the relatively modest dependency of the new migrants and settled agriculturalists.

Those communities with the highest degree of dependency on the forest can obtain all their nutritional requirements from the plants and animals in the forest, as well as fuel, medicines, materials for constructing shelters, tools, weapons and clothing, and can obtain manufactured goods through trading or bartering products such as rattans, gums and resins, *gaharu*, birds nests and many others.

Communities which practice swidden agriculture generally only rely on swidden for their staple diet and supplement it with protein from hunting, fruit from trees cultivated in the abandoned swiddens and in some home gardens, and they also trade and barter many forest products for cash or other manufactured products. Hall (1993) indicates that trade in a number of forest products such as *gaharu* and rattan are very important to the remote communities, although wild price fluctuations and greedy middlemen reduce the direct benefits to the local communities.

The least forest dependent communities in the forested regions are generally the migrants, who have little experience of the rainforest, if they have come from Java. Even they, however, will use poles and other materials from the forest for house construction and may collect fruit and rattans. They are more likely to cultivate rubber on their ladangs, which means a permanent removal of the forest, but they may also become engaged in illegal logging. knowing that there is a big local market, particularly in Java, for timber for construction and furniture making.

Forest Department statistics record 1.26 million households dependent on shifting cultivation on the five main islands, representing about 14% of the rural population of those islands. A major land-use study conducted in the middle 1980s (known as RePPProT) recorded 34.7 million ha on the main islands referred to in this chapter as being shifting cultivation, secondary forest and grassland, which implies an average of 27.5 ha per household. Although lower than the figure of 40 ha per household given by Colfer and Dudley (1992), it may be a more realistic overall figure.

As a crude estimate of the economic value of the forest, it may be assumed that the value of their annual product in terms of the market value of the food and other products obtained from the forest is about US$500 per person, which is less than the current average national income of US$600. This would imply a total annual value obtained from the forest of over US$3 billion per year, or US$90 per ha per annum over the gross area affected.

According to Silitonga (1994), there are around 90 non-forest products currently exported from Indonesia which earned US$350 million in 1989. The most important of these are rattan, gums, oils and resins, but increasing volumes of handicrafts made from wood and other forest products are being exported through tourism. LATIN (1994) lists more than 1250 forest plant species as having medicinal properties. Many of these have formed the basis of Indonesian traditional medicines which has developed into the important Jamu industry, employing numerous people.

Fuelwood and charcoal is still a major energy source in Indonesia, and estimates of the consumption vary from around 60 million to more than 100 million m^3 annually. The highest consumption will be where the population density is greatest and forest area least. Therefore it represents an important economic benefit from the forests least able to meet the supply sustainably since most wood is collected very locally.

Indonesia has set aside some 19 million ha of forests as National Parks and Reserves, of which 5.4 million ha on 31 sites are *in situ* conservation areas of great national and international importance. These areas are being used increasingly for tourism and recreation and can be expected to generate revenues as their management is strengthened and facilities for visitors are developed.

The commercial timber industry remains the dominant component of the forestry sector in terms of income generation. Following the changes in 1967 referred to above, the private sector began operating concessions to extract logs for export, and quickly grew to dominate the world trade in tropical hardwood logs.

As logs became more readily available, the sawmilling industry expanded, and by 1980 around 1650 sawmills were in operation, producing about 4.6 million m^3 of sawn timber. Much of this was concentrated on sawing *ramin* from the swamp forests of eastern Sumatra and western Kalimantan, which provided raw material for the processing of mouldings for export.

In 1980, in an effort to develop processing of forest products within the country, restrictions were placed on log exports, culminating in a total ban in 1985. This led, in particular, to the rapid growth of the plywood manufacturing industry and later to other forms of wood-based panels. Exports of plywood increased from about 280,000 m^3 in 1980 to 8.5 million m^3 in 1990, when heavy export taxes were levied on rough sawn wood to encourage further processing within the country.

The wood processing industry has now stabilised, with an annual production of around 35 million m^3 of logs, and annual exports of sawnwood and panel products combined worth around US$4.5 billion. Recent developments include an expansion of a pulp industry, which will be largely based on plantation grown logs, and some restructuring of the plywood industry to increase value added by additional processing, and bring the industries capacity more in line with the potential future log supply.

The forestry sector is estimated to provide employment directly for more than 300,000 people in logging, forest management, the wood processing industry and government administration. In addition to this there are many small contractors who do not appear in the statistics, as well as the indirect employment in service and other related industries. FAO (1990) estimates that the forestry sector contributes 2.8% to total employment in the formal sector, excluding fuelwood and other non-marketed products.

Forest products represent the second most valuable source of export earnings after oil and natural gas, and visibly contributes more than 2% to gross GDP. The fact that most of the economic activity is in the poorer outer islands, where alternative opportunities are generally limited, increases the overall importance of forestry related activities compared with the hard statistics.

The logging and wood processing industry has generally been very profitable, and the remaining group companies have grown as the number of companies involved has declined due to takeovers and mergers, with several ranking among the largest in Indonesia.

Government revenue from forestry is mainly derived from levies on logs extracted, with a small land rental payment. In addition there is a compulsory contribution to a reforestation fund also based on the volume of logs extracted, which must be deposited in designated bank accounts. The government does receive tax revenues from corporation tax on the profits of the companies too, but it is difficult to determine how much should be attributed to forestry.

The economic value of the forest for commercial logging is best measured by the 'economic rent', which is defined as the 'surplus' from the sale of logs or the products derived from them, from the costs incurred in extraction, transport, forest management, processing, marketing and business management.

Recent studies by a number of agencies (Ramli and Mubariq 1993; Dudung 1992; Tuanakotta and Mustofa 1992) of the economic rent for Indonesian rainforest give a wide range of results, but all conclude that the government only receives a small proportion of the levies and other charges. The companies that operate the concessions receive the largest proportion in the form of profits.

The latest study is based on a series of computer models of a representative concession, sawmill and plywood mill, and these can be used to examine the sensitivity of the economic rent to variations in market prices of products as well as in operating costs. Product prices have a major effect, and probably account for much of the differences between the study results, but operating costs are also important. Some of the differences that are found in operating costs can be attributed to management and general operating efficiency, but a significant proportion is attributable to factors, in particular transport costs, which are location dependant and therefore less amenable to management control.

About 25% of the charges levied by government vary slightly according to the region in which the concession is located, and the species group that is harvested, but the balance, which is the reforestation fund levy is a fixed rate per m^3. Since the regions are very large, costs actually vary greatly within the regions, with the result that profitability varies according to the accessibility and terrain for individual concessions. Companies operating remote concessions with difficult terrain will have lower profitability than those with concessions near the coast or rivers with good forest, and this is likely to act as a disincentive to investing in forest management, and community development in the areas where it is most needed.

The generally low levies and royalties in the past has not only enabled the logging companies to make generous profits, but it has resulted in high levels of waste because the resource is too cheap. Despite this, there is reported to be considerable illegal logging, from which the government receives no revenue. These two factors may contribute to deforestation over time, if the forest is heavily damaged or repeatedly disturbed, and regeneration is inhibited. As the standing stock in the forest declines, pressure to clear the residual forests increases.

5. Developments and land-use planning

In early 1982, after a majority of concessions had been awarded, a process of forest land-use planning was initiated, which is referred to by its Indonesian acronym of TGHK. The task was mainly carried out by the Forestry Department in consultation with the provincial governments, and resulted in a plan, map and area statement for each province. Five broad forest use classes were recognised: production forest, limited production forest (in difficult terrain), protection forest (mainly for watershed protection), conservation forest (consisting of nature reserves, national parks, recreational and tourist areas and other protected areas) and conversion forest.

The plans were drawn up on maps at a scale of 1: 500,000 and endorsed by all government departments concerned with land-use. The result of the exercises was to recognise 142 million ha of the total land area of 192 million ha as 'forest land' which is under the jurisdiction of the Forestry Department. Within this area 30 million ha was classed as conversion forest, and was therefore expected to be cleared at some future date, leaving around 112 million ha of permanent forest, of which about 62 million ha is production forest (including limited production forest), 30 million ha is protection forest, and the balance of 20 million ha is conservation forest.

At the time when this exercise was carried out, there was little information on local site conditions, and the available maps contained little detail and were generally inaccurate. Thus many boundaries could not be identified on the ground and were frequently established in the wrong place. At the same time, resettlement programmes and other projects requiring forest clearances were often established on inappropriate sites because of the lack of information.

Starting in 1984, a regional planning study (RePPProT), was carried out to improve the selection of sites for transmigration projects. This mainly used LANDSAT satellite imagery, and produced a series of natural resource maps at a scale of 1: 250,000 for the whole of Indonesia. These include current land-use maps with forest cover, as well as land systems and land status as established by TGHK.

The improved quality of basic data allowed areas of the TGHK forest land-use classes to be recalculated more accurately, resulting in some minor changes to the figures. However, it also allowed the original TGHK criteria to be applied more rigorously to redraw boundaries, and the current forest cover within each of the classes to be measured. The results showed that current forest cover within the TGHK boundaries was actually only 118 million ha (67% of the area classified), and that the areas of conservation and protection forest should both be substantially greater (25 and 55 million ha respectively), and production forest should be reduced to only 31 million ha in total.

In 1991 the government introduced a new regulation relating to provincial level planning which replaced all or parts of many previous regulations. It provided for a special structure plan (RSTRP) to be drawn up in each province which should determine land-use according to the long-term development needs. The plans were based on the results of land evaluation studies, including the RePPProT study, and made further changes in the boundaries of forest land, and the forest use class within the forest. The proposed changes most affect the production forest areas, as they are also the areas that tend to be best suited to alternative uses.

6.   Forest clearance

The historical changes in forest cover have been described above, and show that forest cover has decreased as population density has increased. Differences in forest cover between provinces in 1982, and between main islands in 1950, show a very similar relationship with population density to the historical trends in each of the main islands.

Most of the reduction in forest area in recent years can be explained by the rapid increases in population density that have taken place, and there does not appear to have been any marked acceleration in the rate of forest clearance which could be attributed to the effect of large-scale logging.

Current rates of deforestation are estimated by FAO (1989) to be around 1 million ha per year, or 1% of the forest area, which is exactly in line with the prediction using the model referred to above. A steady population growth rate of 2% per annum implies an accelerating increase in population density over time as the total population grows, and this could account for the increase in the rate at which forest decreases. Table 3 below shows the impact of a 2% population growth on the percentage cover for an imaginary land area of 500,000 km^2, using an elasticity of forest cover with respect to population pressure of -0.3, from the analysis of the data from Indonesia described above.

TABLE 3. Changes in forest area and in the annual rate of deforestation with a constant population growth rate of 2% per annum

Total land area (km^2)	Population (1000)	Population density (Persons/ km^2)	Years to grow at 2%	Predicted forest cover (%)	Forest area (km^2)	Annual rate of decrease in forest cover (%)
500,000	1,250	25		84	420,000	
	2,500	50	35	54	270,000	1.01
	3,750	75	20	29	145,000	1.03
	5,000	100	14	19	95,000	1.03

A 1% annual decrease in forest area is therefore consistent with the current population densities in the main islands outside Java. On past historical evidence the rate of deforestation can be expected to increase slightly as populations continue to increase.

Commercial logging is frequently blamed for causing deforestation, and it may be a contributory factor through improving access, but its overall effect would appear to be small in relation to the needs of a growing population for land to grow food and other crops.

As mentioned above, the proportion of the rural population which is currently practising traditional shifting cultivation is only 14% of the total or 1.26 million households, but the FAO (1989) study attributes more than half the annual deforestation to shifting cultivation. Various studies of shifting cultivation suggest that the average area cleared annually ranges between 1 and 2.5 ha per household (Colfer and Dudley 1992; Hall 1993) with the proportion in old-growth forest declining from about 75% in recently established communities to only 15-20% in old established communities, where travel distances have become large. The estimates of 500,000 ha cleared annually by shifting cultivators is therefore consistent with these figures.

Traditional shifting cultivators allow the forest to fallow after cropping, and will recut it after 15-20 years, but increasingly new settlers are clearing forest for an annual crop, and then planting perennial crops such as rubber. Sometimes this represents a form of land speculation, and results in permanent loss of forest. Frequently the perennial crops are of very poor quality and low yielding. There is a danger that the area under crops such as rubber will expand beyond that needed to meet demand, prices will drop and many smallholder crops will no longer be economic. To reduce wasteful destruction of the forest, more intensive cultivation of better quality crops should be encouraged, and guidance needs to be given to farmers on the future market prospects for crops.

Farmers who clear forest for food production and cultivation of perennial crops receive an immediate and direct return on their investment of time, but receive little or no direct benefit from the forest, which goes mostly to the wood using industry and government. Extension services could help to improve the returns for farmers from agriculture and crops, but in addition a share of the benefits from forest management could act as an incentive to be less wasteful when clearing the forest.

7. Political considerations

The main beneficiaries from forest clearance are those that clear it, since they generally do not pay for the land as most of the forest is on state land and the forest is a national resource. Some communities claim traditional rights to their land, and although such rights are recognised to exist, they are subordinated under the law to the state's rights as custodians of the land. Very few, if any, claims to traditional rights have been investigated and registered, with the result that there is always great uncertainty about the validity of claims.

Although there are procedures for claiming and registering title to land through the National Agrarian Board, this tends to be slow and expensive for low income farmers. Sometimes awarding of land titles is linked to transmigration schemes where individual lots of 1-2 ha may be titled. However, for traditional shifting cultivators, who may require a large area for rotating their swiddens, their only claim is through local traditional rights, which are increasingly difficult to protect in the face of pressure on land from newcomers to the area.

Responsibility for land-use planning and local development lies with the provincial governments, and priorities locally for activities which require land, such as plantations, transmigration schemes, mining developments and industrial development may differ from those of central government. Thus, the argument for clearing forest for a project may look very different from the perspectives of the provincial and central government.

The provincial governments receive 45% of the proceeds of the forest levies, of which one third is passed down to the *bupati* second tier local government. The current levies vary according to species group and the region that the logs come from. A few rare and special species such as sandalwood and ebony are charged a high levy, but on meranti, the main commercial species group, levies range between US$ 8.40 and 10.40. So, provincial governments only receive US$ 4-5 per m^3, which is very small compared with the central governments' share of the levies plus the US$ 16 per m^3 which must be paid into the Reforestation Fund, and represents only about US$ 4-5 per ha at present yields.

In forest rich provinces, this revenue may provide a small incentive to the provincial governments to support measures to avoid excessive forest clearance. However, the rental value of the forest captured by government is small compared with many alternatives and reductions in the forest area do not immediately show up as reduced revenue, since the levies are based on the volume felled, and not the area of forest. Several provinces are exporters of logs and therefore derive no benefit from the downstream processing investment and employment and may have little incentive to resist measures to convert forest to other uses which generate more revenue.

8.   Recent measures to establish permanent forest

The discussion on land-use planning and forest clearance above has shown that, at least until recently, the latter has proceeded without regard for the former. The most recent planning exercise has attempted to take current land-use into account, but the data base is now 10 years old, and changes have continued throughout that time. This situation arises because the planning is done at a macro level, and cannot discriminate boundaries in sufficient detail to accommodate all the small annual changes that are taking place.

In order to overcome this problem the Forestry Department decided in 1991 to create new management units throughout the production forest areas on the main islands based on long experience in Java. These units are referred to locally as Kesatuan Pengusahaan Hutan Produksi (KPHP). The decree which provided for the establishment of KPHPs sets out general principles, and the details are now being developed with a number of pilot units in different provinces. The size of the units is flexible according to the purpose of the management, local forest and site conditions, and the economics of managing the unit. A unit can therefore be natural forest, plantation forest or a mixture of the two, and it can vary in size according to whether it is primarily to be managed for industrial wood production or to meet local community needs.

The three most essential differences from what has gone before are: wherever possible clear natural boundaries will be used; the general pattern and distribution of the units is set out for a province using the latest structure plans; and the process of establishment uses a 'bottom up' rather than 'top down' approach for the detailed boundaries, which must be surveyed and negotiated on the ground, with the participation of local communities. It is hoped that this will avoid the problems of the past, where communities have often been ignorant of the forest boundaries defined by the government, and have felt free to clear forest wherever they may choose, and the units will be managed rather than being covered by a permit to harvest timber, as with the present concessions.

This does not mean that communities will no longer be able to clear forest, but rather that areas to be cleared in the foreseeable future should be discussed and agreed in advance so that the boundaries and the forest within them really are permanent. Only when forest areas are permanent can sustainable forest management be practised, as continual erosion of the forest area means that harvesting cycles cannot be maintained, and production should decline, or stop all together if an area due to be harvested after a period is no longer forest.

Within the pilot KPHPs a complete baseline inventory has been carried out, which has four functions: to enable the condition of the forest to be monitored and compared at periods in the future for assessing sustainability, to provide an estimate of the current overall growth of the forest as a basis for determining the allowable annual harvest, to provide the necessary information on the forest for preparing a detailed long-term management plan which is held in a Geographic Information System (GIS): and to provide information on the extent and distribution of non-timber forest products (NTFP) which could be managed and harvested by local communities to improve their income.

This approach differs from that used in current concessions which only inventories that part of the forest to be harvested in the coming 20 year period. The management plan for the KPHP is also expected to differ from those in current use, in that areas within the forest which might be suited for some non-productive purpose such as conservation, tourism of protection can be identified and marked on the maps. It will also make provision to integrate plans for harvesting of timber with those of NTFPs to avoid any interference but achieve any mutual benefits that may exist, such as access.

Since 1992, five pilot KPHPs have been in process, and the general pattern of all the proposed KPHPs in three provinces and in a district in two other provinces has been delineated. Recently a number of concession holders have requested to establish KPHPs where their concessions are consistent with the general pattern. The Forestry Department plans to create 195 KPHPs within the current five year development plan period to 1999, but this may prove optimistic if only by a few years.

Although the KPHP provides the mechanism for sustainable forest management, it will only be possible to say that sustainability has been achieved after several repeat inventories have been carried out at 10 year intervals. However, the availability of a GIS and the information it contains will make it much easier to make judgements.

9.   Future prospects for Indonesian forests

As discussed above, the forest area in Indonesia is now thought to be about 100 million ha. However, on the basis of the land classification, almost 80 million ha of this should be protection forest or for conservation and recreation. These areas will be particularly difficult to protect because they are the responsibility of provincial or central government with limited manpower and financial resources, and they generally do not generate income to cover the cost of management and protection.

A number of projects have been established in the past few years to investigate the possibility of local communities playing an active role in management and protection of reserves. These have all involved considerable financial resources as well as expertise, which has so far come mainly from international sources. The results reported to date are encouraging, but the question remains as to where the resources will come from to extend the ideas countrywide within a reasonable time frame.

With the production forest being licensed to commercial operators there are better prospects. The companies are employing increasing numbers of professionally trained staff who live and work in the forest. Some of the companies are showing signs of being willing to become forest managers rather than just users of the forest, and the KPHP will push them in that direction.

For some time companies have been increasingly aware of potential future shortages of logs and have been establishing plantations to make up the anticipated shortfall. The advantages of plantations are their high yields of single species, which means that the volume output from 1 ha of plantations may be ten or twenty times that from natural forest. On the otherhand they require heavy investment of capital, carry substantial risks, and since 'fast growing' species are normally used, they may have limited uses.

Long-term plans are for around 4 million ha of industrial plantations with annual programmes of about 500,000 ha. The total area currently of timber plantations is around 800,000 ha and a realistic figure likely to be about 2 million ha in the foreseeable future. Even this reduced area has the potential to produce as much as or more than currently comes from natural forest. However, a majority of plantations are being developed to produce raw material for a pulp and paper industry, for which they are ideally suited, and so the existing plywood industry will have to continue to depend on the natural forest.

There are already signs that the plywood industry is beginning to contract, but the prospects are not entirely gloomy, because recent inventories and sample plots have shown that if the natural forest is left undisturbed after logging, it recovers very well, and there seem to be very good prospects for reharvesting after 35 years in accordance with present guidelines. The concession holders generally harvest only a limited number of species, typically 15-30 from 200 plus, and so with better knowledge about the volumes and size class distribution of species in the forest from the KPHP inventories, there are prospects to develop markets for currently unused species where their timber has properties that suit a particular use. Promotion of 'lesser known' species has often yielded disappointing results in the past because of lack of knowledge on the potential supply.

With most of Indonesia's forests in the relatively sparsely populated islands, it is difficult to find the human resources to introduce new measures rapidly, but encouraging progress has been made over the past five years, which must give grounds for cautious optimism for the future.

10.   Conclusions

Deforestation in Indonesia has been taking place for a long time (cf. Maloney, this volume), but the observed increase in the rate in recent years seems to be very largely explained by rapid population growth in the heavily forested islands outside Java. It can therefore be expected to continue at current rates until population stabilises, unless changes in demographic factors occur. These could be a growth in urbanisation, which is presently very low in Indonesia, greater intensification of agriculture or changes in patterns of internal migration.

The five main forested islands are unlikely ever to reach the existing population density on Java, which still has 20% forest cover, but Sumatra and Sulawesi appear likely to lose forest cover until it decreases to around 40% if current trends continue. Kalimantan and Maluku seem likely to finish with nearer 60% and Irian Jaya with 80% forest cover. In overall terms these would mean a forest cover nationally of about 60 million ha in contrast to the 100 million ha at present.

At a current deforestation rate of about 1 million ha per year this would take 40 years, or one generation. It is impossible to forecast what other changes may take place over that period, especially intensification of agriculture, as has happened in Java and in Europe, and it is possible that long before that time the trend could be reversed.

The negative side to such a *laissez-faire* outlook is that once destroyed the natural rainforest can never be replaced. The results of reafforestation are usually monotonous plantation monocultures. It is therefore essential that efforts are stepped up to involve local communities in local land-use planning, so that unnecessary and wasteful forest clearance is avoided.

The prospects for a slightly slimmed down wood based industry using a mixture of plantation grown wood and logs from the natural forest would seem to be fair, and with better resource data and planning new products can be developed to add value in order to be less dependent on volume for profitability.

References

Beccari, O. (1904). *Wanderings in the Great Forests of Borneo*, Constable, London.

Bock, J.P. (1881). *The Head-hunters of Borneo*, Low, Marston and Rivington, London.

Colfer, C.J.P. and Dudley, R.J. (1992). *Shifting Cultivators in Indonesia*, FAO, Rome.

Dudung, D. (1992). Economic rent from forest utilisation. Paper presented at the Seminar on Economic Rent, Jakarta, 6-7 October 1992, Forestry Department/APHI, Jakarta.

FAO (1990). *Situation and Outlook of the Forestry Sector in Indonesia*, Vol. 2. *Forest Resource Base*, Ministry of Forestry/FAO, Jakarta.

Hall, J.C. (1993). *Managing the Tropical Rainforests: Swiddens, Housegardens and Trade in Central Kalimantan*, Oxford Brookes University, Oxford.

Hugo, G.J., Hull, T.H., Hull, V.J. and Jones, G.W. (1987). *The Demographic Dimension in Indonesian Development*. East Asian Social Science Monograph, Oxford University Press: Singapore.

LATIN (1994). *Sustainable Production of the Diversity of Medicinal Drugs from Indonesia's Tropical Forest*, Forestry Faculty, IPB and Lembaga Alam Tropika Indonesia (LATIN), Bogor.

Lennertz, R. and Panzer, K.F. (1984). Preliminary Assessment of the Drought and Forest Fire Damage in East Kalimantan, Report of fact-finding mission, German Agency fro Technical Cooperation (GTZ), Jakarta.

Ponder, H.W. (1982). Land-use in new and old areas of Iban settlement, *Borneo Research Bulletin*, **14**, 3-14.

Ramli, R. and Mubarriq, A. (1993). *Forest Utilisation Economic Rent in Indonesia*, WAHLI,Jakarta.

Repetto, R. and Gillis, M. (1988). *Public Policies and the Misuse of Forest Resources, Cambridge* University Press, Cambridge.

Silitonga, T. (1994). *Non-wood Forest Products of Indonesia*, in P.B. Durst, W. Ulrich and M. Kashio (eds.), *Non-wood Forest Products in Asia*, FAO, Bangkok.

Sutter, H. (1989). *Forest Resources and Land-use in Indonesia*, Ministry of Forestry/FAO, Jakarta.

Tillema, H.F. (1938). *A Journey Among the Peoples of Central Borneo in Word and Picture*, Van Munsters Vitgevers, Amsterdam.

Tuanakotta, H. and Mustofa (1992). *Wood Economic Rent for Commercial Logs; data for the year 1991*, Association of Forest Concession Holders of Indonesia, Jakarta.

Wallace, A.R. (1869). *The Malay Archipelago*, Macmillan, London.`

Dr. Alastair Fraser, UK-Indonesia Tropical Forest Management Project, Manggala Wanabakti Building, Jalan Gatot Subroto, Jakarta 10270, Indonesia

8. DIVERSITY DESTROYED?
The monoculture of Eucalyptus

Christopher J. Barrow

1. Introduction

Due to human activities, it is increasingly rare for natural regeneration of tropical forest to occur after disturbance and it is seldom easy to restore, with human assistance, something close to the original cover. Usually biodiversity-depauperate secondary forests, scrubland, grassland, farmland, pasture or planted forests are left. The greater the area disturbed and the greater the disturbance (Simons 1988: 46), the less likely is there to be satisfactory recovery (for a discussion of the barriers to regrowth and restoration techniques tried in Amazonia, see Nepstad *et al.* 1991).

Although not a precise definition, reafforestation can mean:

1. an attempt to restore something approaching natural cover (artificial regeneration);

2. planting a different type of forest on previously planted ground: either stands of one or a few species or enrichment planting of selected species among the remaining forest (reafforestation);

3. establishment of forest on 'previously unforested' land, which probably means (Burgess 1993; Mather 1993: 2) loss rather than gain of biodiversity (afforestation).

Plantation forestry can be reafforestation or afforestation and generally results in stands of a limited number of tree species and a management regime that restricts the biodiversity associated with the planted trees. If plantations can meet timber demand or provide a livelihood for people there might be less encroachment on natural forests and greater preservation of biodiversity. However, there are also potential side-effects: plantation species may escape and compete with existing cover elsewhere, plantations may burn and kill the wildlife they harbour more easily than natural forest, they may expose surrounding forests to greater fire risk, or they may alter the quantity and quality of runoff and affect groundwater levels. Plantations are thus both a threat and a benefit and may affect biodiversity well beyond their immediate vicinity. Unfortunately, reafforestation means the replacement of rich-biodiversity natural vegetation with poor-biodiversity tree cover. It is possible that the high esteem given to eucalypt planting has helped discourage replanting of clear-felled areas or enrichment planting with indigenous species.

There are few tropical countries where forests and woodlands are not being destroyed faster than they are regenerating or replanted (Anon. 1986a). Plantation area estimates are available (Lanley and Clement 1979; Evans 1986a, 1992). Figures

B.K. Maloney (ed.), Human Activities and the Tropical Rainforest, 151-168.

published by Burgess (1993: 136) suggested that annually for every (roughly) 17 ha cleared (roughly) 2.6 ha were planted.   Mather (1993: 4) suggested the area of plantations was likely to be between 125 and 150 million ha (ca. 3 to 4% of global forest area). Figures for 1990 show that about 43 million ha was in the tropics and planting there seems to have accelerated since then (Evans 1992).

If the world's remaining forests are to be conserved a key factor will be the satisfaction of demand for timber products from some other source(s). The growth of eucalypts or other suitable rapid-growth trees on degraded or unforested land (e.g. the vast areas of species-depauperate *Imperata* grasslands in Southeast Asia) may help prevent more deforestation {although in practice (Evans 1992: 33), eucalyptus and other fast-growth tree plantations have often caused cutting of natural forest}.

2.   Eucalypts

Eucalypts have been a popular plantation species for decades, however, there are still knowledge gaps and controversy about the use of the genus. There are over 600 species. of eucalyptus growing in a wide range of habitats: lowland humid tropical, tropical and subtropical semi-arid, tropical and subtropical high altitudes and other frost-prone regions (Pryor 1976; FAO 1981; Evans 1986b) and roughly 30 of these are in general use.

During the 1970s and early-1980s eucalypts were widely seen as 'miracle' trees: fast growing, adaptive and offering other advantages. The Tropical Forestry Action Plan and, before the late-1980s, the World Bank encouraged and supported the spread of eucalypt plantations. By the late-1980s over 4 million hectares of eucalypts (Anon. 1986a) had been planted outside the natural range of the genus (Australia, Southeast Asia and the Pacific).

Development has fashions and fixations with what are perhaps limited, inappropriate, approaches to problems. In plantation forestry it has been the use of eucalypts and alternatives may have been neglected in their favour. In India, for example, where eucalypts are a key part of social forestry, it has been claimed that less information is available about the qualities of native species than eucalyptus, and that much of the information on the latter is too favourably biased (Budd *et al.* 1990: 311; Shiva 1993: 31-39).

Plantation forestry has mainly focused on maximisation of yield but there are other goals which deserve attention: sustainability, appropriateness, quality/appropriateness of product, employment generation, amenity and conservation value, etc. Development activities are often difficult to assess and efforts to do so lead to false-impressions and over-generalisations. Attempts to assess the value of eucalypt planting must consider the trees and their context. Problems may be more a reflection of poorly designed social forestry programmes or the pressure of market forces, rather than unsuitable trees (Joyce 1988; Sharma 1993).

Pulp production has been expanding in warmer developing countries, probably because tree growth is faster and, thus, profits greater (Sunder and Parameswarappa 1989), and perhaps because land can often be appropriated from the poor or has no title-holder. As species diversity of natural forests tends to be greater in lower latitudes there is a relatively greater threat to biodiversity if plantations replace natural vegetation there, rather than in higher latitudes. Eucalypt planting looks set to expand. Therefore it is important that unwanted physical and socio-economic impacts should be minimised while benefits must be maximised. There is also a need to assess the way eucalypts are grown and alternatives to the planting of eucalypts.

3.   Questions about eucalypts which need consideration

The following questions concerning planting of eucalypts need to be taken into account:

1.   Which species should be grown?

Much botanical information has been gathered, if not adequately disseminated, (e.g. about moisture needs, growth-form, timber qualities, value as fuelwood, productivity, soil and climate tolerance, vulnerabilty to pests, speed of growth under good conditions, etc) but practical experience needs to be collected and made accessible.

2.   What form should planting take?

Should large areas be planted, or small on-farm plots, planting along roadsides, agroforestry, or should trees be planted alongside (Saxena 1991) or amongst crops (Igboanugo and Otu 1988; Kirk *et al.* 1990: 29)?

3.   How should planting be organised?

Should individuals or the state plant? Should larger or smaller farmers have support? Should common land be used? Should planters own the land? What market will the trees serve?

4.   Are eucalypts the best species to grow or should they be grown with or replaced by others?

This will need careful assessment of a wide range of on-site and off-site costs and benefits, including factors like long-term climatic change.

5.   Can generalisations be made reliably from local level experiences?

Given the number of available species, and the range of local physical and socio-economic conditions this may be unwise or difficult (a point made by Conroy 1992: 25). Perceptions vary and groups in a society may suffer to a differing degree while local practice may also reduce or magnify various impacts. It might be possible to develop some sort of expert system which can be interrogated before planting proceeds.

The Swedish International Development Authority (SIDA) undertook an assessment of the ecological impacts of eucalypts for the FAO (cf. Poore and Fries 1985) which is based on those studies; Anon. 1986a).

4.   The use of eucalypts: recognised advantages and disadvantages

The recognised benefits and of planting eucalypts are:

      1.   They have fast enough growth to make investment attractive.

Under most economic conditions slow-growing trees are unlikely to prove attractive to private investors while eucalypts give a reasonably rapid return, plus an interim profit from thinnings and governments have generally supported and subsidised plantations. In contrast, they have neglected to fund reafforestation with slow-growing native species.

      Fast growth is vital where fuelwood demand exceeds supply: it offers good possibilities for commercial profit, may interest industrialised countries seeking 'sink' plantations to offset their carbon dioxide emissions (ideally in tropical, developing country, environments for rapid growth and lower investment costs). However, some people doubt that eucalypts really are rapid growers in less favourable environments, and feel that local tree species might offer as good or better growth rates and, may have other advantages, such as nitrogen fixation, production of better quality fuelwood, etc (Shiva 1993: 31-39). Rapid growth may be negated if timber or fuelwood is somehow inferior or if the established trees fail to prevent erosion or cause other problems.

      2.   There are eucalypt species able to grow in virtually all conditions, temperate to tropical, low to high altitude, humid to semi-arid.

However, only a small portion of the total have been 'tested' for plantation use, and it is fair to say that a few species are over-promoted.

      3.   Good field experience has been assembled for some species.

These species were cultivated in over 80 countries by 1986 but it has been argued that planting them hinders a consideration of other eucalypts and non-eucalypt taxa.

      4.   Some eucalypts can be coppiced and this results in rapid production of straight poles and considerable biomass

      5.   Eucalypt plantations tend to yield straight trees which are attractive to commercial producers.

      6.   It is claimed that eucalypt plantations are a form of 'sustainable' forestry.

However, few plantations have been established long enough to see if repeated cropping damages soil and the claim is hardly ecologically sound if plantings displace valuable natural vegetation (Hurst 1990: 191; Durand 1993).

7.   There is a possibility of increasing eucalypt productivity through genetic improvement.

8.   Eucalypts provide a fuelwood source.

There are situations where eucalypt planting has proved vital for fuelwood supply (e.g. in parts of Ethiopia and the Andes). Whether other species would have performed as well or better is generally unknown. The benefits may be more than simple satisfaction of fuel needs. Nesmith (1991) assessed the effects of planting in west Bengal and concluded the time taken by women gathering fuel had fallen from up to 8 hours a day 4 days a week to 2 hours a day 4 days a week. This time saving might have health and welfare advantages, assuming the beneficiaries used the time saved for useful tasks. If they do prove to be a successful fuelwood source planting of eucalypts may allow more dung to be used on the land (Conroy 1992). However, at least some eucalyptus spp. wood is fast-burning, consequently plantations may not ease fuelwood shortages and prices may be misleadingly cheap (people need to burn more). The quality as fuel may be poor (some eucalypts emit unpleasant aromatic smoke making them unsuitable for cooking, smoking food or for use in open hearths). Where the wood is usable the risk is it will be sold away from the area where there is a wood shortage or at prices too high for the poor to afford. The solution is to support local planting and ensure the poor have access to the wood (in some parts of India lower castes use the leaves for fuel and this can lead to soil erosion).

9.   Eucalypts yield a wide range of non-fuel products.

For instance, eucalypts yield timber, leaves (fuel for poor), oils: some species (e.g. *Eucalyptus citrodora*, *E. globulus*, *E. logirostris*) support essential oil/gum production. The oils derived from eucalyptus spp. are: cineol, citronella, citronellol, piperitone, etc {for a list of oils and producer species (see FAO 1981: 294-296; Kirk *et al.* 1990: 23)}. The wood can be used for pulping, woodchip feedstock, veneering, or as tanbark and some species are nectareous, so apiculture is possible, yielding honey, beeswax and royal-jelly. In the Northern Hemisphere eucalypts are early-flowering which allows expansion of apiculture.

10.   Eucalypts are highly productive, so less land may be needed to meet the timber demand.

This could help take pressure off of natural forests. Unfortunately in practice there have often been cases where the opposite happens and natural forest is cleared to plant eucalypts.

11.   Eucalypts provide little canopy shading so they may allow agroforestry or grazing beneath (certainly for the first 4 or 5 years,

perhaps not for the remaining 4 or so of a typical rotation) or dense planting of trees to get more and straighter growth.

A thin canopy is pointless if leaves poison crops nearby/below and/or roots extract soil moisture denying crops or forage. Lack of cover and ground vegetation may allow severe soil erosion, especially if leaves are collected.

12.   Plantations can be vertically integrated.

They support steel-making (Brazil, Argentina and Australia), the making of paper products, rayon, etc., and provide associated employment, foreign earnings, etc., but they can also lead to factory pollution of air and streams.

13.   Eucalypts can help rehabilitate degraded land or yield produce from wasteland.

This could be one of the greatest benefits of planting eucalypts. There are many examples of drought-tolerant eucalypts being effectively used to establish dryland fuelwood lots and shelterbelts (Kohli *et al.* 1990) and barriers to moving sand dunes, especially for protecting roads (Goudie 1990: 42, 192; Mather 1993: 96). Because eucalypts are unattractive to browsing animals (even goats) they are useful for fringe-planting to protect other trees or crops (Alexander,1989) while salt-tolerant species may be used in salinized/waterlogged areas to provide wood and to counter waterlogging/salinization by drawing-down groundwater through their evapo-transpiration (Haas 1993: Schofield 1992).

The disadvantage of planting eucalypts are that:

1.   Some people have been marginalised by eucalypt planting, losing common land and suffering reduced employment opportunity.

Eucalypts have become a cause of grievance and symbol of 'top-down' (authoritarian) planning (Kardell *et al.* 1986; Joyce 1988; Conroy 1992) in some countries. There can be problems if eucalypts replace commons fodder, fuelwood, grazing and other usufruct practices, or if planting allows large landowners to avoid employing staff. If common lands are planted with eucalypts for pulping, etc., people may have to seek fuelwood elsewhere. Weighing planting benefits against costs may not be easy as some common land is well-managed and capable of sustained use, while other common land might be over-grazed and, if left unplanted with eucalypts, become eroded and useless within a few years (Anon. 1986b; Shiva and Bandyopadhyay 1987: 19). There are also widespread reports of lost employment, but in some cases there may be improved opportunities, at least for some groups.

2.   Monocropping leaves the trees potentially vulnerabile to pests, diseases or environmental changes.

However, eucalypts are widely planted outside their natural range and have so far had relatively few insect or fungal problems, although the introduction of *Eucalyptus* in South Africa and New Zealand did lead to the spread of insect pests (Elton 1977: 110).

3.   Eucalypts are unattractive to browsing animals.

In fact this can be beneficial as it makes them easier to establish where there is uncontrolled grazing, which is useful if the trees then provide erosion control or fuelwood or act as windbreaks or form a belt to guard less robust trees/crops.

4.   Eucalypts may have a negative effect on nearby crops or pasture due to moisture competition and allopathic effects

Some eucalypts exude compounds which damage or kill nearby crops or grasses and inhibit seed germination (allopathic effect). It is not clear whether such effects continue for long after a plantation has been removed.

5.   Eucalypts may upset established drought mitigation strategies.

If eucalypts displace trees which offer dry season/drought fodder or food, there may be an erosion of people's drought-coping strategies.

6.   Short-rotation tree cropping probably depletes soil nutrients.

Eucalypts are felt by many to be likely to degrade soil (Anon. 1986a: 22). However, there are, dissenting voices (e.g. Kirk *et al.* 1990). Typically, eucalypts are grown on a 8 to 14-year rotation (often with thinnings before the final cut), and are often coppiced. Repeated cropping may reduce soil nutrients, especially if litter is collected for fuel. The 'solution' may be to ensure litter and bark are left at cutting site and to grow legume crops or nitrogen-fixing trees (e.g. *Sesbania sesban*, *Casuarina* spp. or *Albizia arborea*) near or in rotation with eucalypts.

7.   Erosion occurs in eucalypt plantations.

The problem seems worst where eucalypt stands are dense and there is little or no groundcover (Poore and Fries 1985). Collection of leaf-litter for fuel adds to the problem. Much depends on how the trees are planted, how the plantation is managed, whether or not fires occur, the type of soil and the climate; it is difficult to generalise.

8.   Eucalypt plantations are unattractive as a habitat/roost for wildlife.

The degree of biodiversity loss partly depends on length of rotation and method of planting and management. Hard facts are difficult to come by and it is worth note that the widely-held idea that conifer plantations are poor in wildlife has recently been challenged. Commercial producers are likely to be more interested maximising uniform, marketable biomass from eucalyptus plantations than conserving biodiversity. This is an advantage in that eucalypts tend not to harbour pests which attack surrounding crops.

9.   Eucalypts do not bear attractive fruit and may offer an unattractive environment to wildlife.

It is possible to manage fringes of plantation blocks or fire-breaks to make the environment more attractive to wildlife and tourists.

10.   Eucalypt plantations may be more fire-prone than the natural vegetation that they replaced.

Where plantations penetrate natural forests in more humid regions they may allow fires to spread where there would otherwise have been much less risk.

5.   Regional experiences

5.1.   BRAZIL (INCLUDING AMAZONIA)

Mather (1993: 5) cites figures suggesting the area of plantations in Brazil increased from 0.5 to 7.15 million ha between 1965 and 1990, and may reach 12 million ha by AD 2000. The first plantings were made in Brazil's centre-south by the Paulista Railway Company in 1910. More recently there have been extensive plantings (of mainly *E. grandis*) notably in Bahia and Minas Gera'is (north east Brazil). In the 1960s Aracruz Celulose SA, with Brazilian Government support, took over a considerable area of land in Espirito Santo and Bahia which had been degraded by charcoal burners, loggers, farmers and ranchers and established huge plantations (Sargeant and Bass 1992: 90). It is very unlikely that anything like the original forest cover could have regenerated on this land if it had not been developed. These Aracruz plantations are grown mainly to produce pulp on a 7-year rotation. The Company has made provisions for conserving blocks of native forest amongst the plantations and claims that planting has not removed natural forest, that it has rehabilitated degraded land, that people have not been marginalised, that disease risk associated with monocultures is reduced by planting a diversity of varieties of eucalypts, and that there has been creation of employment.

In Bahia State considerable areas of the remaining original tropical Atlantic forest (*mata atlantica*) have been cleared to plant eucalypts. The main use (Joyce 1988) is for charcoal to support the steel industry (which gets over 40% of its energy from eucalypts). Not only has natural forest been displaced, largely to make charcoal, but the production and use of that fuel is also leading to air pollution and acid deposition which is likely to have serious impacts on flora, fauna, soils and water bodies. Considerable areas have been planted in Para (eastern Amazonia) to provide charcoal for steel smelting along the Caraj'as - Sao Louis Railway (Fearnside 1989).

There have been extensive plantings of *E. deglupta* at Jar'i (Plate 1) in eastern Amazonia to provide cheap paper pulp but there have been problems with soil compaction and erosion where these and other plantation species have been grown. Some of the difficulties may have been the result of management more by civil

PLATE 1. Jar'i Agroforestry Project (Brazilian Amazonia). In the 1970s an American entrepreneur (Daniel Ludwig) purchased *c.* 1.4 million hectares of land from the Brazilian govenment on the Rio Jari, eastern Amazonia.   Ludwig calculated that there was a growing market for paper pulp and that the planting of exotic, fast-growing species in Amazonia's hot, humid environment offered potential.   The pulp factory (background) was towed by barge from Japan and was to be supplied by around 200,000 hectares of *Eucalyptus deglupta*, *Gmelina arborea* and *Pinus coriba via* a 200km railway network (middle picture). Planting began in the late 1970s but by the time this photograph was taken (1981) only 110,000 ha was established and growth rates were much slower than had been expected.   Not only had natural forest been cleared, but the pulp factory was being fed with logs cut from surrounding forests to make up for plantation deficiencies. Owned by a consortium of 22 Brazilain companies since 1982, it was still not possible to assess whether or not it was a success in 1990.

engineers than foresters. The main impacts have been the loss of forest areas cleared for planting and air/river pollution associated with pulp production (Fearnside 1989). Growth rates at Jar'i have been reported to be disappointing.

5.2.   PERU

Many Andean valleys have been cleared of natural forest and the occupants now rely on eucalypt planting for fuel and building timber (mainly *E. globulus*). Planting takes place at 3000 m or more (Whitmore 1987). This planting might help to protect lower altitude tropical forest if it meets fuelwood demands.

5.3.   CAMEROON

Eucalypts are popular with small farmers in Cameroon.

5.4.   ETHIOPIA

Since 1895, when a French engineer suggested that it should be introduced from Kenya, *E. globulus* has been about the only tree species widely planted in Ethiopia's Central Highlands. There are now around 100,000 ha planted (Leach and Mearns 1988) wherever rainfall is over 800 to 1000 mm yr^{-1} (some at 2600 m altitude or higher). The trees are mainly coppiced and used for fuelwood and poles. Planting is popular with small farmers and seems to have caused fewer problems than in India or Thailand, possibly because there is no large pulp company involvement (Pohjonen and Pukkala 1990, 1991). Without eucalypts the fuel situation would have been desperate long ago, natural tree cover may have suffered for the worse and dung and crop residue would have been burnt more often, leading to greater soil degradation and probably more land clearance for subsistence agriculture (indeed, the capital Addis Ababa, moved every few years before eucalypts were planted, as a consequence of fuelwood shortage). There have been problems of moisture removal and claims the eucalypts naturalise and endanger native species survival, however, the latter fear may be misplaced because in most of the areas planted the indigenous flora had already been decimated (Joyce 1988).

5.5.   KENYA

Eucalyptus camaldulensis is widely grown as a cash-crop on farm woodlots to provide poles and fuelwood (as a cash-crop). Planting has reduced pressure on natural vegetation for fuelwood and has almost certainly reduced land degradation and soil erosion (Joyce 1988). However, there have been worries that *E. camaldulenis* is more prone to forest fires than natural vegetation. Poor farmers as well as richer seem to have enjoyed benefits (Chambers and Leach 1989; Chambers *et al.* 1993; Nesmith 1994), and it is not unusual for income derived from marketing the trees to be used to pay school fees, medical bills, dowries, provide for old age, etc (such investment in trees is not restricted to eucalypts and is widespread throughout the world). Kenyan tobacco growers plant eucalypts to provide fuel for curing their crop.

5.6.   MALAGASY REPUBLIC

A good deal of the vast area of natural forests that have been degraded has been replanted with eucalypts in the Malagasy Republic, but it may be that some of this planting is helping to 'drive' deforestation.

5.7.   NIGERIA

Eucalyptus camaldulensis plantations are used to reclaim tin-mine spoil found to increase soil acidity (Anon. 1988; Alexander 1990) in Nigeria. In savanna regions eucalypts may offer a means of providing fuel and building timber, helping protect natural tropical savanna tree cover (Adegbehin 1990; Igboanugo *et al.* 1990).

5.8.   UGANDA

Evans (1992: 339) reported some success with using *E. robusta* plantings to dry-out papyrus swamps.

5.9.   RWANDA

There is a long history of eucalypt planting and the tree is popular with small growers. Some old plantations are now moribund and have suffered soil erosion and there is a trend to replant with pine species (Kerkhoff 1990: 20).

5.10.   INDIA

Originally introduced in 1843 to meet fuelwood demands, eucalyptus spread rapidly between 1980 and 1985, after which prices fell in many areas and a many growers lost confidence. The bulk of planting has taken place in Gujarat, west Bengal, Orissa, Mahdya Pradesh, Bihar, Kerala and Karnataka {where Chandrashekhar *et al.* (1987) reported problems as a consequence of eucalypt planting}. *E. terecornis* is the most widely grown species (and is often referred to erroneously as 'hybrid' eucalyptus}, but *E. globulus* and *E. camaldulensis* are also common (for a bibliography, see Shiva and Bandyopadhyay 1987).

Larger landowners in most areas have done well, finding a market for poles or, more often, selling to pulp/woodchip producers. For these people eucalypts can mean employing much less labour, compared to landowners planting a crop like, cotton. The low labour requirement and unpalitability to livestock means that farmers may take other employment and need hire little or no labour to cover while they are away (Ascher and Healy 1990: 55-60; Saxena 1992a).   There may well be situations where employment opportunities in towns has depleted the rural work force so that the opportunity offered by eucalyptus is welcome (Joyce 1988: 59). For some smallholders eucalyptus growing may allow opportunities to accumulate savings or pay-off debts (Chambers and Leach 1989; Chambers *et al.* 1993). Eucalypts can offer a safe income or form of capital requiring little labour and farmers can then work in the cities or as wage labourers for other farmers (Davidson 1983; Shah 1984). For others, especially the landless and those who need resources from common land, eucalypt planting can spell increased hardship. Not only may the support for and the impact of eucalypt planting vary amongst different social class groups in a society, but it is also likely to differ between men and women as the latter are often less likely to own land and thus benefit (Nesmith 1994: 140-143).

In late-1970 the National Social Forestry Project was started (receiving World Bank and British ODA funding). The aim of this was to encourage farmers to plant for their subsistence needs (fuelwood and construction timber). By the early-1980s there had been a good deal of planting (especially in southwestern West Bengal), but most for the market (mainly for pulp and poles) rather than subsistence (possibly because fuelwood commanded too low a value compared with the aforementioned alternatives). Thus programme benefits were effectively 'hijacked' by larger landowners who profited because labour costs are less than for traditional crops. Possibly as much as 2.5 million ha (mainly in Gujarat, Uttar Pradesh and eastern Rajasthan) came under farm forestry between 1981 and 1988 and over two-thirds of this was for eucalypts (Saxena 1992b: 54).

Eucalypts declined in popularity after 1986, this trend was reinforced when the Indian Government realized that eucalyptus was becoming a cash-crop and adopted a new Forest Policy (in 1988) which removed many of the subsidies that had been paid to growers. Saxena (1992b) attributed the dissatisfaction to:

1. production problems which cut expected yields (poor seedlings, poor soils and inadequate weeding, planting too densely);

2. supply/demand problems (pulp mills were not keen to take smaller growers wood, and sales for fuelwood were made through middlemen who controlled prices and/or farmers were ignorant of realistic market prices);

3. market inadequacies (in some states legal problems prevented farmers from selling fuelwood directly);

4. negative impact on agricultural production (the production of some crops declined, leading to food shortage and price rises);

5. farmers found that crops near eucalypts suffered (cf. Shiva and Bandyopadhyay 1983; Baldwin and Bandhu 1990; Kirk *et al.* 1990: 8; Conroy 1993).

Saxena (1992c) observed that in Uttar Pradesh and a few other states, labour shortages had helped encourage farmers to plant eucalyptus, and smaller farms with less labour problems found the trees less attractive.

In some regions, for a variety of reasons (Saxena 1992d), farm forestry with eucalypts has been a success. Conroy (1993: 2) discovered that in some areas of India, where the laws allowed, small farmers found out timber merchants prices, charged slightly less and avoided the middlemen to make good profits. Whether or not supplies of eucalypt wood have helped reduce cutting of natural tree cover is difficult to establish. Government spending on promoting alternative energy supplies like kerosene or solar power might have been more effective than expenditure on eucalypt planting.

In 1983 the Indian Government initiated the Karnataka Social Forestry Programme, in part as a response to rapid deforestation of natural tree cover. Common (*panchayat*) land was allocated for eucalypts with the result that many poor were evicted and faced a falling demand for their labour. The eucalypts went for paper/rayon pulp or sale for poles, so the fuelwood supply situation got worse, at least for the poor who no longer had access to common land or waste from annual crops (Budd *et al.* 1990: 313) while converting the land back to native trees or annual crops may prove difficult.

As in Thailand, considerable peasant opposition to eucalypt planting has developed. Conroy (1992: 19) noted ..."the real social problem in Karnataka was not eucalyptus *per se*, but the fact that ownership of the common land on which they were grown was transferred to the pulp and paper industries...".

5.11.   INDONESIA

Natural forests have been heavily exploited for woodchip and pulp production in Indonesia, and have often been replaced with eucalypt plantations.

5.12.   CHINA

For a review see Song (1992).

5.13.   CAMBODIA

Considerable planting of *Eucalyptus camaldulensis* has been undertaken in Cambodia.

5.14.   THAILAND

Thailand has suffered rapid and serious loss of natural forest. In the early part of this century there was probably about 75% tree cover, while in 1961 only about 53% of the land area was tree-covered, in 1988 just 18%, and by 1991 approximately 15% (Lohmann 1989; 1991b). Considerable areas have been planted with eucalypts since 1978 (mainly *E. camalmulensis*). Lohmann (1991a) estimated 8% of Thailand's area had a cover of eucalypts by 1990. Most planting has been by larger growers (some being joint-venture Thai/MNC, Thai/Japanese or Thai/Taiwanese) producing pulp/woodchip for export to paper and rayon manufacturers, especially in Japan and S. Korea. Lohmann (1991a: 81) blamed Thai and foreign businesses for the loss of natural forest ('economics-driven loss of biodiversity'), noting that before the 1970s they were intent on logging, later on pulp production. It is claimed that the Royal Thai Forestry Department has been more active in promoting the planting of eucalypts than in conserving natural forest but in the long term the Thai Government would like to convert some of the plantations to teak or other trees (Hurst 1990: 227).

Roughly 50% of eucalypts in Thailand in 1992 were privately-owned, most had been planted after 1988, a large number in the northeastern region (Patanapongsa 1990; Puntasen *et al.* 1992).

Considerable tax incentives were offered to eucalypt growers by the Thai government from the early-1990s. Logging bans have also helped prompt eucalypt planting (Lohmann 1989). About 80% of Thailand is without legal title but people traditionally enjoy usufructory rights (Puntasen *et al.* 1992: 196; Sargeant and Bass 1992: 23). The people using untitled land can be treated as 'squatters' and be moved off by eucalyptus growers, while those with usufruct rights may be prompted to sell usufruct rights to business, often through middlemen who make large profits. Having sold rights people then clear forest reserve land and re-start the process of speculation, which has serious impacts on the natural vegetation (Lohmann 1990; 1991a). Puntasen *et al.* (1992: 204) commented that: "The introduction of eucalypts to Thailand was meant to be for reafforestation ...unfortunately... it has been switched to a 'cash' crop". The loss of access to common land and, real or imagined, environmental impacts, have over the last 5 years or so led to a 'resistance movement', and in some areas trees have been grubbed-up by angry peasants. Lohmann (1991b: 10-15) argues that this commercialisation of common land and undermining of the base of local subsistence is a new manifestation of enclosure, and that ultimately 500,000 families might be displaced. Whether or not as a consequence people are now more conscious of the need for forest conservation is difficult to judge. Even if they are, their marginalisation may have made it more difficult for them to support conservation.

5.15.   THE PHILIPPINES

Eucalyptus deglupta is widely grown in the Philippines, either as plantations or for enrichment planting in logged forest which is not regenerating fast enough to satisfy exploiters (especially in Mindanao). The Paper Industry Corporation of the Philippines (PICOP) promotes eucalypt planting and is probably the biggest wood and paper pulp producer in Asia. PICOP delegates much of the management to smaller growers, issuing 'stewardship certificates'. It has large plantations but the planting is often by the smallholders whom it co-ordinates. PICOP tries to encourage mixed planting of eucalypt, crops and pasture, but in practice most growers just plant blocks of trees. Typically growers get seedlings at cost price, cultivation advice, up to 75% loans to pay for establishment and a contract to buy the wood at a negotiated price with the freedom to sell elsewhere if need be (Sargeant and Bass 1992: 117). There is little indication in the literature, as is the case for most countries, of whether eucalypts spread from plantations to become a nuisance alien species.

6. Conclusion

Eucalypt species are grown in a wide range of tropical environments, from semi-arid to humid tropical, at low altitude and in upland areas. Planting is often in the form of single-species blocks which may offer limited opportunities for wildlife conservation. The reputation eucalypts have for offering opportunities for quick profits encourages their cultivation and has often led to losses of common land, including natural forest areas, reduced employment opportunities and even cutbacks in local food production. Some authorities condemn the use of eucalypts and in more than one country poorer sections of the community have identified the trees as symbols of oppression and exploitation. There is also a widespread fear that eucalypt planting damages the soil and water resources.

It is clear from the literature that it is difficult to generalise reliably about eucalypts: local circumstances vary and must be allowed for. Ideally planting should take place on degraded land and not lead to loss of natural forest or good croplands. Socio-economic realities may make such allowances difficult. The benefits of conserving biological diversity have seldom been reaped by local people (Jusoff and Majid 1995), whereas richer land owners and companies can profit from establishing eucalypts.

Ensuring eucalypt planting protects, rather than hastens, the destruction of biodiversity and does not disadvantage local livelihoods will probably require state intervention, incentives for local people, and careful consideration of subsidies for planting. There are a wide diversity of eucalypts, so it should be possible to find species that flourish in almost any but the most extreme environments, and which have enough resource value to be attractive. Planted on the right land eucalypts can provide wood and other products and therefore reduce pressures on natural forests. There is a need to move from assessing growing needs and growth characteristics to developing more sensitive planting strategies that exploit the strengths of eucalypts and avoids their destructive effects.

Note

1.   exotic = a species which has been introduced or which has spread to an area where previously it was not naturally found.

References

Adegbehin, J.O. (1990). Establishment and management of eucalyptus plantation in some parts of the savanna region of Nigeria, *Savanna*, **11** (2), 35-44.

Alexander, M.J. (1989). The long-term effect of Eucalyptus plantations on tin-mine spoil and its implications for reclamation, *Landscape and Urban Planning*, **17** (1), 47-60.

Alexander, M.J. (1990). Reclamation after tin mining on the Jos Plateau, Nigeria, *Geographical Journal*, **156** (1), 44-50.

Anon. (1986a). Are eucalypts ecologically harmful?, *Unasylva*, **58** (152), 19-22.

Anon. (1986b). Giving India's trees to the people, *The Economist*, **301** (7475), 93-94.

Anon. (1988). Is agriculture a viable alternative to eucalyptus plantations on reclaimed tin-mine spoil on the Jos
    Plateau, Nigeria?, *Environmental Conservation*, **15** (3), 261.

Ascher, W. and Healey, R. (1990). *Natural Resource Policymaking in Developing Countries: environment,
    economic growth, and income distribution*, Duke University Press, Durham, North Carolina.

Baldwin, J.H. and Bandhu, D. (1990). Social forestry in Karnataka State: India, in W.W. Budd, I. Duchard,
    L.H. Hardesty and H. Steiner (eds.), *Planning for Agroforestry*, Elsevier, Amsterdam, pp. 293-320.

Bandyopadhyay, J. and Shiva, V. (1985). *Eucalyptus* in rainfed farm forestry: prescription for
    desertification. *Economic and Political Weekly*, **20** (40), 1687-1688.

Budd, W.W., Duchard, I., Hardesty, L.H. and Sieiner, H. (eds.) (1990). *Planning for Agroforestry*, Elsevier,
    Amsterdam.

Burgess, J.C. (1993). Timber production, timber trade and tropical deforestation, *Ambio*, **22** (2‑3), 136-143.

Calder, I.R., Hall, R.L. and Prasanna, K.T. (1993). Hydrological impact of eucalyptus plantations in India,
    Journal of Hydrology, **150** (2-4), 635-648.

Chambers, R. and Leach, M. (1989). Trees as savings for the rural poor, *World Development*, **17** (3), 329-342.

Chambers, R., Leach, M. and Conroy, C. (1993). *Trees as Savings and Security for the Rural Poor* (Gatekeeper
    Series No. SA3), Earthscan, London.

Chandrashekar, D.M., Krishnamurti, B.V. and Ramaswamy, R.S. (1987). Social forestry in Karnataka: an
    impact analysis, *Economic and Political Weekly*, **22** (24), 935-941.

Conroy, C. (1992). Can eucalyptus be appropriate for small farmers?, *Appropriate Technology*, **19** (1), 22-25.

Conroy, C. (1993). Eucalyptus sales by small farmers in eastern Gujarat, *Agroforestry Systems*, **23** (1), 1-10.

Davidson, J. (1983). *Setting Aside the Idea that Eucalypts are Always Bad*, FAO, Rome.

Durand, F. (1993). De la sylve aux plantations d'eucalyptus, 25 ans de gestion forestiere en Indonesie (1967-
    1992), *Archipel*, No. **46**, 191-217.

Elton, C.S. (1977). *The Ecology of Invasions: by animals and plants* (1st edn. 1958), Chapman and Hall,
    London.

Evans, J. (1986a). Plantation forestry in the tropics - trends and prospects, *International Tree Crops Journal*, 4,
    3-15.

Evans, J. (1986b). Assessment of cold-hardy eucalypts in GB, *Forestry*, **59** (2), 223-242.

Evans, J. (1992). *Plantation Forestry in the Tropics: tree planting for industrial, social, environmental and
    agroforestry purposes* (2nd edn.), Clarendon Press, Oxford.

FAO (1981). *Eucalyptus for Planting* (2nd edn.), FAO Forestry Series No. 11, FAO, Rome.

Fearnside, P. (1989). The charcoal of Caraj'as: a threat to the forests of Brazil, *Ambio*, **13** (2), 141-143.

Goudie, A.S. (ed.) (1990). *Techniques for Desert Reclamation*, Wiley, Chichester.

Haas, J. (1993). Effect of saline irrigation on early growth of *Eucalyptus gomhocephala* and *Acacia saligna*,
    Environmental Conservation, **20** (2), 143-148 and 162.

Herwitz, S.R. and Gutterman, Y. (1990). Biomass production and transpiration in eucalypts in the Negev
    Desert, *Forest Ecology and Management*, **31** (1-2), 81-90.

Hurst, P. (1990). *Rainforest Politics: ecological destruction in South-East Asia,* Zed Books, London.

Igboanugo, A.B.I., Omijeh, J.E and Adegbehin, J.O. (1990). Pasture floristic composition in different *Eucalyptus* species plantations in some parts of northern Guinea savanna zone of Nigeria, *Agroforestry Systems,* **12** (3), 257-268.

Igboanugo, A.B.I. and Otu, M.F. (1988). Economic implications of eucalyptus as agroforestry trees - an overview, *Savanna,* **9** (2), 24-28.

Joyce, C. (1988). The tree that caused a riot, *New Scientist,* **117** (1600), 54-59.

Jusoff, K. and Majid, M.M. (1995). Integrating needs of the local community to conserve forest biodiversity in the State of Kelantan, *Biodiversity and Conservation,* **4**, 108-114.

Kardell, L., Steen, E. and Fabio, A. (1986). Eucalyptus in Portugal - a threat or promise?, *Ambio,* **15** (1), 6-13.

Kerkhof, P. (1990). *Agroforestry in Africa: a survey of project experience*, PANOS, London.

Kirk, C.F., Dury, S.J., Judd, N. and Axlerod, P.G. (1990). *Eucalyptus for Plantation Forestry: benefits, concerns and comparisons: a review*, Commonwealth Forestry Institute, Oxford.

Kohli, R.K., Singh, D. and Verma, R.C. (1990). Influence of eucalypt shelterbelt on winter season agroecosystems, *Agricultural Ecosystems and Environment,* **3** (1), 23-31.

Lanley, J.P. and Clement, J. (1979). *Present and Future Natural Forest and Plantation Areas in the Tropics*, FAO, Rome.

Leach, G. and Mearns, R. (1989). *Beyond the Woodfuel Crisis: People, Land and Trees in Africa*, Earthscan, London.

Lohmann, L. (1989). Forestry in Thailand: the logging ban and its consequences, *The Ecologist,* **1** (2), 76-77.

Lohmann, L. (1990). Commercial tree planting in Thailand: deforestation by any other name, *The Ecologist,* **2** (1), 9-17.

Lohmann, L. (1991a). Who defends biological diversity? Conservation strategies and the case of Thailand, in V. Shiva., P. Anderson, H. Schacking, L. Gray, L. Lohmann and D. Cooper (eds.), *Biodiversity: social and ecological perspectives*, Zed Books, London, pp. 77-104.

Lohmann, L. (1991b). Peasants, plantations and pulp - the politics of eucalyptus in Thailand, *Bulletin of Concerned Asian Scholars,* **2** (4), 3-18.

Malik, R.S. and Sharma, S.K. (1990). Moisture extraction and crop yield as a function of distance from a row of *Eucalyptus tereticornis, Agroforestry Systems,* **12** (2), 187-195.

Mather, A. (ed.) (1993). *Afforestation: policies, planning and progress*, Belhaven, London.

Nepstad, D.C., Uhl, C. and Serrao, E.A.S. (1991). Recuperation of a degraded Amazonian landscape: forest recovery and agricultural restoration, *Ambio,* **20** (6), 248-255.

Nesmith, C. (1991). Gender, trees, and social forestry in West Bengal, India, *Human Organization,* **50** (4), 337-348.

Nesmith, C. (1994). Trees for rural development, *Applied Geography,* 14(2), 135-152.

Pantanapongsa, N. (1990). Private forestry development in Thailand: a survey of tree growers in the northeast region, *Commonwealth Forestry Review,* **69** (1), 63-68.

Pohjonen, V. and Pukkala, T. (1990). *Eucalyptus globulus* in Ethiopian forestry, *Forest Ecology and Management,* **36** (1), 19-31.

Pohjonen, V. and Pukkala, T. (1991). Which eucalypt grows best in Ethiopian Highlands, *Biomass and Bioenergy,* **1** (4), 193-195.

Poore, M.E.D. and Fries, C. (1985). *The Ecological Effects of Eucalyptus*, FAO Technical Papers/Forestry Papers No. 59, FAO, Rome.

Pryor, L.D. (1976). *Biology of Eucalypts*, Arnold, London.

Puntasen, A., Siriprachai, S. and Punyasavatsut, C. (1992). Political economy of eucalyptus: business, bureaucracy and the Thai Government, *Journal of Contemporary Asia*, **22** (2), 187-206.

Sargeant, C. and Bass, S. (eds.) (1992). *Plantation Politics: forest plantations in development*, Earthscan, London.

Saxena, N.C. (1991). Crop losses and their economic implications due to growing eucalyptus on field bunds - a pilot study, *Agroforestry Systems*, **16** (3), 231-245.

Saxena, N.C. (1992a). Adoption of a long-gestation crop - eucalyptus growers in north-west India, *Journal of Agricultural Economics*, **43** (2), 257-267.

Saxena, N.C. (1992b). Eucalyptus planting as a response to farm management problems faced by 'on-site' and 'off-site' farmers, *Agroforestry Systems*, **19** (2), 159-172.

Saxena, N.C. (1992c). Eucalyptus farmland in India: what went wrong?, *Unasylva*, **43** (170), 53-58.

Saxena, N.C. (1992d). Farm forestry and land-use in India: some policy issues, *Ambio*, **21** (6), 420-425.

Schofield, N.J. (1992). Planting for dryland salinity control in Australia, *Agroforestry Systems*, **20** (1-2), 1-23.

Shah, S.A. (1984). The case for eucalyptus, *Indian Express Magazine*, 15/7/84, 1,6.

Sharma, R.A. (1993). Socioeconomic evaluation of social forestry policy in India, *Ambio*, **22** (4), 219-224.

Shiva, V and Bandyopadhyay, J. (1983). Eucalyptus - a disastrous tree for India, *The Ecologist*, **13** (5), 184-187.

Shiva, V and Bandyopadhyay, J. (1987). *Ecological Audit of Eucalyptus Cultivation* (Research Foundation for Science and Ecology, Dehra Dun), Sagar, New Delhi.

Shiva, V. (1993). *Monocultures of the Mind: perspectives on biodiversity and biotechnology*, Zed Books, London.

Simons, P. (1988). Costa Rica's forests are reborn, *New Scientist*, **120** (1635), 43-47.

Song, Y.F. (1992). Utilization of eucalypts in China, *Appita Journal*, **45** (6), 382-283.

Sunder, S.S. and Parameswarappa, S. (1989). Social forestry and eucalyptus, *Economic and Political Weekly*, **24** (1), 51-52.

Whitmore, J.L. (1987). Plantation forestry in the tropics of Latin America: a research agenda, *Unasylva*, **39** (156), 36-41.

Dr. Christopher J. Barrow, Centre for Development Studies, University of Wales Swansea, Swansea SA2 8PP, Wales. U.K.

9. MODELLING CLIMATIC IMPACTS OF FUTURE RAINFOREST DESTRUCTION

K. McGuffie, A. Henderson-Sellers and H. Zhang

1. Modelling tropical deforestation

Tropical forests provide the habitats of about half of the world's species and are an important natural sink of CO_2. Since strongly ascending branches of the Walker and Hadley circulation are located over tropical forest regions, changing the land-surface characteristics by removing the rainforest may affect the atmospheric circulation. For these reasons, the impacts of tropical deforestation on the local, regional and global climate have received considerable attention from climate analysts and modellers in the last decade (e.g. Henderson-Sellers and Gornitz 1984; Dickinson and Henderson-Sellers 1988; Nobre et al. 1991; Henderson-Sellers et al. 1993; McGuffie et al. 1995; Sud et al. 1996). The existing global climate model (GCM) simulations of tropical deforestation offer some general agreement about the likely impacts on the local and regional climate, despite the considerable differences in the GCM dynamical structures, land-surface representation, ocean description and length of the simulations (Table 1).

Among the significant properties of tropical rainforests are that they have a very low surface albedo throughout the year, their leaf area and stem area are larger than those of any other vegetation and the trees are tall. Replacing the tropical rainforest with grassland leads to three primary changes in the land-surface properties (Figure 1):

i. the surface albedo is increased, which directly causes a reduction in the land-surface net radiation;

ii. the reductions in the leaf area and stem area lead to a decrease in the water holding capacity of the vegetation and thus reduce the re-evaporation of the intercepted precipitation and, probably, the vegetation transpiration;

iii. the grassland replacing the tropical forest is much shorter and smoother than the rainforest so that the surface roughness is dramatically reduced and the surface frictional forcing is weakened.

The decreased surface roughness length, which is the primary factor in determining the aerodynamic exchange between the land surface and the lower atmosphere, has two complementary effects on the change of evapotranspiration. Even though the strengthened surface wind speed mitigates the effects of the reduction in surface roughness, the net effect of these two processes seems to be such that the surface evapotranspiration is decreased. The change in surface evapotranspiration, acting in connection between the changes in the hydrological processes (determining the regional water recycling) and the changes in the land surface and atmospheric energy budget (the sink of the surface energy budget and the source of the atmospheric energy budget), has a crucial role in determining the local impact of tropical deforestation.

Uncertainties still exist in the results of climate simulations with respect to the magnitude and direction of changes, however (e.g. Table 1). In trying to explain the reasons for these differences, some authors emphasise the importance of surface albedo

B.K. Maloney (ed.), Human Activities and the Tropical Rainforest, 169-193.
© 1998 *Kluwer Academic Publishers. Printed in the Netherlands.*

TABLE 1. Results of climate model simulations of the impacts of tropical deforestation over the Amazon Basin

Reference (1)	D & H-S 1988	L & W 1989	Nobre et al. 1991	L & R 1992	D & K 1992	Dirmeyer 1992	H & S et al. 1993b	P & L 1994	Sud et al. 1995	McG et al. 1995	Zhang et al. 1995a	Zhang et al. 1996
GCM	CCM OB (4.5X 7.5)	UKMO (2.5X 3.7)	NMC (1.8X 2.8)	UKMO (2.5X 3.7)	CCM1 -Oz (4.5X 7.5)	NMC (4.5X 7.5)	CCM1 -Oz (4.5X7 .5)	LMD (2.0X 5.6)	GLA (4.0X 5.0)	CCM1- Oz (4.5X 7.5)	CCM1- Oz (4.5X 7.5)	BMRC (2.0X 3.0)
Ocean	Fixed SST	Fixed SST	Fixed SST	Fixed SST	Slab Ocean	Fixed SST	Slab Ocean	Fixed SST	Fixed SST	Slab Ocean	Slab Ocean	Slab Ocean
Surface	BATS	Canopy	SiB	Canopy	BATS 1e	SSiB	BATS 1e	SECHI BA	SSiB	BATS1e	BATS1e	BEST
Integr- ation Length (2)	!/3 year	3/3 year	1/1 year	3/3 year	3/3 year	14/14 mth	6/6 year	11/11 year	3/3 year	14/6 year	25/11 year	10/6 year
Albedo Change	0.12 to 0.19	0.136 to 0.188	0.12- 0.14 to 0.16- 0.24	0.136 to 0.188	0.12 to 0.19	incre ment of 0.03	0.12 to 0.19	0.135 to 0.216	0.097 to 0.146	0.12 to 0.19	0.12 to 0.19	0.15 to 0.21
Rough- ness Change	2.0 to 0.05	0.79 to 0.04	2.65 to 0.08	2.0 to 0.05	2.0 to 0.05	2.65 to 0.08	2.0 to 0.05	2.3 to 0.06	2.65 to 0.077	2.0 to 0.2	2.0 to 0.2	1.1 to 0.1
ΔT (K)	+3	+2.4	+2	2	0.6 (soil)	Not given	+0.6	-0.1	+1.3	+0.3	+0.3	+0.9
ΔP (mm)	0	-490	-640	-295	-511	+33	-588	-186	-266	-437	-402	+445
ΔE (mm)	-200	-310	-500	-200	-255	-146	-232	-128	-350	-231	-222	+248
ΔE-ΔP	-200	+180	+140	+95	+256	-179	+356	+58	-84	+205	+180	-197

1. D & H-S (1988) is Dickson and Henderson (1988); L & W (1989) is Lean and Warrilow (1989); L & R (1992) is Lean and Rowntree (1992); D & K (1992) is Dickinson and Kennedy (1992); H-S et al. (1993b) is Henderson-Sellers et al. (1993b); P &L (1994) is Polcher and Laval (1994); McG et al. (1995) is McGuffie et al. (1995).

2. Lengths of integration are represented by control/deforestation. All the changes are annual averages over the Amazon Basin

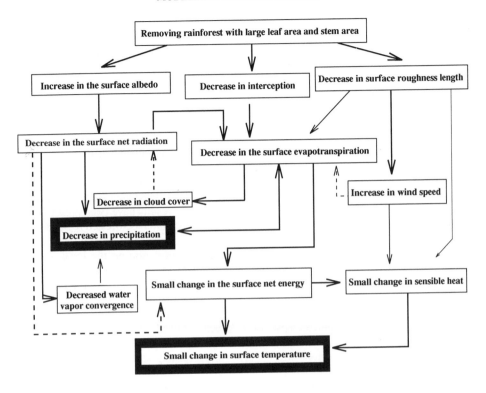

FIGURE 1. Schematic illustration of processes occurring in the atmospheric column
and affecting surface temperatures after deforestation.

in terms of the Charney mechanism (Charney 1975; Mylne and Rowntree 1992; Dirmeyer 1992), while others stress the role of changes in surface roughness length and soil water capacity (Mintz 1982; Cunnington and Rowntree 1986; Dickinson and Henderson-Sellers 1988; and Henderson-Sellers *et al.* 1993). One particular shortcoming in most model simulations of the impact of rainforest destruction is the relatively short length of the integrations (Table 1), which, as argued in Henderson-Sellers *et al.* (1993), probably does not allow the climate system to achieve a new equilibrium after the disturbances are imposed.

Recent simulations by Polcher and Laval (1994a,b) and McGuffie *et al.* (1995) confirmed that over different regions, the impacts of deforestation vary in magnitude and mechanism, an effect also proposed by Mylne and Rowntree (1992). Results from Henderson-Sellers *et al.* (1993) suggested that there were teleconnections between the two deforested regions via the Walker circulation. These results have been confirmed by a more recent study (Zhang *et al.* 1996b) which considered three deforested regions.

In this chapter, the impacts of the tropical deforestation on the local, regional and global scale climate system are illustrated by using results from a version of the National Center for Atmospheric Research (NCAR) Community Climate Model (CCM1) incorporating the Biosphere-Atmosphere Transfer Scheme (BATS1e) (Williamson *et al.* 1987; Dickinson *et al.* 1986, 1993) henceforth termed CCM1-Oz. The experiment discussed here consists of a 25-year 'control' integration and an 11-year 'deforestation' integration in which the tropical forests in the Amazon Basin, Southeast Asia and tropical African regions are converted into a scrub grassland.

The deforestation scenario used here is the same as that in McGuffie *et al.* (1995) and Zhang *et al.* (1996a,b) but differs from most of the previous simulations in terms of the location of deforestation, it includes tropical Africa, and the characterization of the deforested landscape. Figure 2 shows the rainforest regions modified in South America, Southeast Asia and Africa in this simulation. There are eighteen grid points of tropical forest in South America, nine grid points in Southeast Asia and eight points in tropical Africa in the grid-point space of CCM1 which has a rhomboidal truncation at wave number 15 (about 4.5° latitude x 7.5° longitude). All the thirty five grid points of tropical forest are 'deforested': replaced by a land type characteristic of a tall grass-covered scrubland with a few large trees and a partial understorey. The major changes in the land-surface parameters are an increase of surface albedo, a reduction in surface roughness length, a decrease in leaf area and changed soil conditions (Henderson-Sellers *et al.* 1993). The exact nature of this characterization could be improved by drawing on detailed field studies (e.g. Giambelluca 1996).

Section 2 assesses the local climatic changes simulated as a result of the imposed tropical deforestation. In Section 3, the effects of tropical deforestation on the regional moisture dynamics are examined. This leads to the evaluation, in Section 4, of possible mechanisms by which the removal of tropical forests can modify the climates of the mid and high latitudes. Section 5 presents a summary and overview of these results.

2.   Impacts of tropical deforestation on the local and regional climates

The simulated impacts of tropical deforestation on the local climate system focuses on three regions: the Amazon Basin from 15°S to 5°N and from 80°W to 50°W, Southeast Asia from 15°S to 20°N and from 95°E to 150°E and tropical Africa from 10°S to 10°N and from 10°W to 30°E (Figure 2).

2.1   IMPACTS ON THE SEASONAL CYCLE

Climatic changes are evaluated over the full seasonal cycle, because alterations in seasonality may have important effects on ecosystems. Figueroa and Nobre (1990) and Nobre et al. (1991) suggested, for example, that if the annual total precipitation remains the same after deforestation, but the dry season is lengthened and has lower rainfall, with slightly greater rainfall occurring in the rainy season, local ecosystems could be affected in spite of small or zero changes in the annual averages.

Seasonal climatic variations before (solid lines) and after (dashed lines) deforestation over the Amazon Basin, Southeast Asia and Africa are shown in Figure 3, which also includes the results of a Student's t test represented by a P-value (the possibility that changes are not statistically significant) which is arbitrarily cut off at 25% in these plots. P-values greater than 5% are not significant. Thus only months exhibiting a black (P-value) bar shorter than 5% (right hand scale) may indicate an impact of deforestation which is statistically significant and hence worthy of further investigation.

Deforestation causes a reduction in monthly total precipitation throughout the year with larger changes during the rainy season: the decrease in DJF in the Amazon (Figure 3a) is almost three times as large as that in JJA although the percentage changes in precipitation during the dry season (JJA) and rainy season (DJF) are nearly the same (around 20%). The pattern in Southeast Asia is similar but the changes are not statistically significant. The largest amounts of precipitation in tropical Africa (Figure 3g) occur during March-April-May and September-October-November. In August-September-October monthly precipitation shows a statistically significant reduction of between 15 and 30 mm month⁻¹ but there are also increases in precipitation in March and April, the former being statistically significant. Deforestation may, therefore, as discussed by Nobre et al. (1991), substantially reduce the possibility of forest regrowth because of the increased length of the dry season. In addition, much lower rainfall in the dry season could increase the possibility of fire which could retard the regrowth of secondary forest.

Statistically significant reductions in evaporation are stimulated throughout the year in all three deforested regions (Figures 3b, e and h). The much smaller reduction in evapotranspiration than in total precipitation suggests that there is a decrease in the large-scale moisture convergence as also reported in most of the earlier model simulations (e.g. Lean and Warrilow 1989; Nobre et al. 1991; Henderson-Sellers et al. 1993; McGuffie et al. 1995; Sud et al. 1996).

The changes in surface temperature differ from region to region and between the warmer and cooler seasons (Figures 3c, f and i). In Southeast Asia, there are small

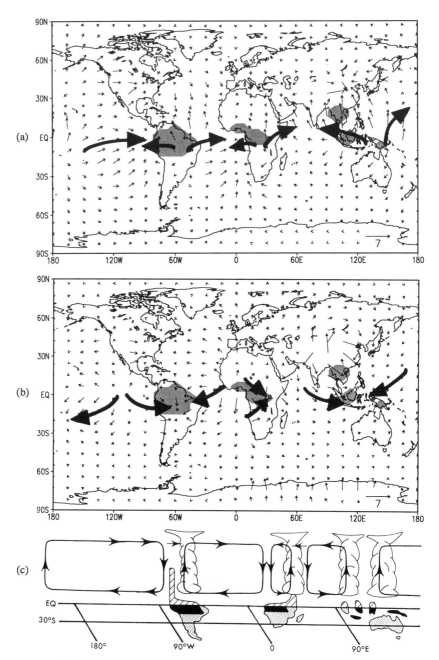

FIGURE 2. Regions where tropical rainforest is removed and atmospheric circulation
systems that might be affected (a) divergent airflow at 200hPa (ms^{-1}) in
January as simulated by CCM1-Oz. Shaded areas are the location of tropical
rainforest in this model resolution and the schematic arrows represent the
features of the upper level tropical atmospheric circulation; (b) as (a) but at
800hPa; (c) schematic diagram showing the position of tropical rainforest in
the vertical structure of the tropical Walker Circulation.

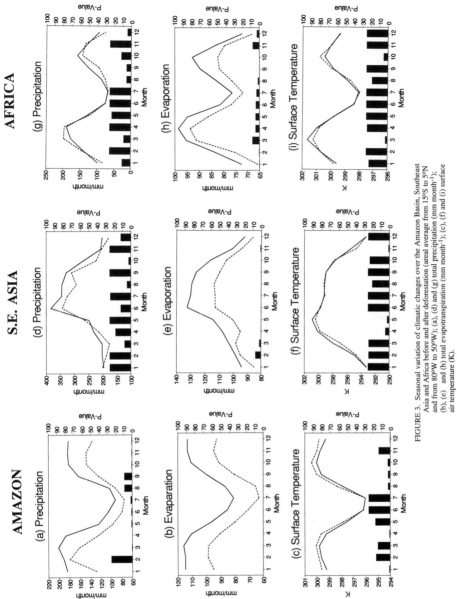

FIGURE 3. Seasonal variation of climatic changes over the Amazon Basin, Southeast Asia and Africa before and after deforestation (areal average from 15°S to 5°N and from 80°W to 50°W); (a), (d) and (g) total precipitation (mm month[-1]); (b), (e) and (h) total evapotranspiration (mm month[-1]); (c), (f) and (i) surface air temperature (K).

decreases in surface temperature (Figure 3i) in most months, and the decreases during the period from February to May (the maximum reduction is about 1° C in March) are larger than those in the period of the Northern Hemisphere summer monsoon when the surface temperature is affected very little by deforestation. Only the changes in April-May and November are statistically significant which suggests that during the transition from the summer monsoon to the winter monsoon the land surface thermal conditions are most easily influenced by deforestation.

2.2   IMPACTS ON THE DIURNAL CYCLE

Assessing changes in the diurnal cycle can help to improve understanding of the impacts of tropical deforestation and their possible consequences for natural and agricultural ecologies. For example, Dirmeyer (1992) suggested that decreased cloudiness during the day-time and increased cloudiness during the night-time almost offsetted the effects of the imposed increase in surface albedo and resulted in the rather small reduction in total precipitation that they found. However, because of the huge size of the datasets which are needed to analyse the diurnal cycle, only a few papers reporting GCM simulations have included this kind of analysis (e.g. Nobre et al. 1991; Dirmeyer 1992).

Here the diurnal cycle is studied by using a sample of dates from the middle of the two months: 15-22 January and July. In each case, values from nine grid points are averaged to reduce the effects of spatial and temporal sampling. Figure 4 shows changes in diurnal cycles over nine central Amazon Basin points as time averages from the 15th to the 22nd of January (Figure 4a-d) and of July (Figure 4e-h).

In January (the Amazonian wet season) the diurnal variations of surface temperature are enhanced following deforestation due to cooling during the night and warming during the day-time (Figure 4a). These changes can be explained in terms of the increase in incoming solar radiation and the reduction of the surface evapotranspiration (Figures 4c, d). The reduction in cloud cover following deforestation is responsible for the increase in solar radiation, even though the surface albedo is increased. Observations over the deforested regions in Amazonia by Bastable et al. (1993) support the results reported here. Bastable et al. (1993) found that the temperature range over the deforested region was twice that over areas of undisturbed forest.

The maximum rainfall (~10 mm d^{-1}) occurs during the night and early morning (Figure 4b), but the surface runoff (not shown here) exhibits quite clear diurnal variation with large surface runoff in the late afternoon, presumably as a result of the strong convective precipitation. The simulated diurnal cycles would therefore appear to be consistent with observations and with other simulated results (e.g. Randall et al. 1991; Hendon and Woodberry 1993). After deforestation, precipitation and surface runoff are weakened, indicating that afternoon convection is affected.

In the Amazonian dry season (July), deforestation results in increases in air temperature, similar to the impacts seen in the wet season (Figure 4e). The reduction of day-time evapotranspiration and the small increase of incident solar radiation following deforestation dominates the cooling effect of increased surface albedo to produce a warmer surface (Figures 4h, g). The diurnal patterns of rainfall and surface runoff are quite similar to those in the wet season with both being decreased after deforestation (Figure 4f). Enhanced low level winds and the reduction in low level specific humidity

JAN   JUL

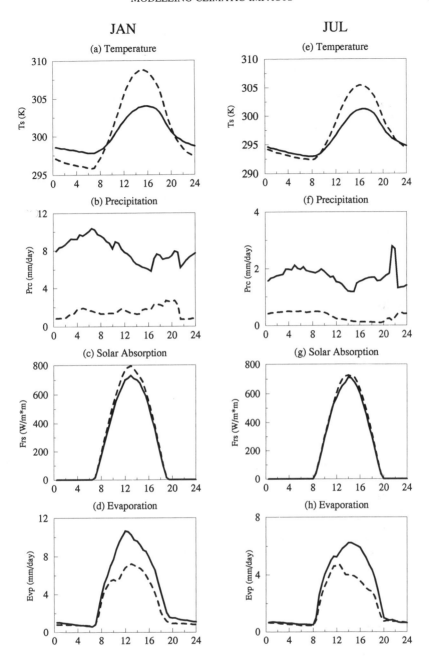

FIGURE 4. Diurnal cycle (average of 9 GCM points) over the central Amazon Basin
for control (solid line) and deforestation (dashed line): time averaged from 15-22
January or 15-22 July: (a) and (e) temperature of the air above the canopy (K);
(b) and (f) precipitation (mm d^{-1}); (c) and (g) net absorbed solar flux (W m^{-2});
(d) and (h) evapotranspiration (mm d^{-1}).

are also seen (but not shown here) in July as a result of the changes in low level dynamics.

2.3   IMPACTS ON THE ENERGY BUDGET OF RAINFOREST REMOVAL

Atmospheric circulation is dependent upon the surface energy budget and the atmospheric energy budget and changes in these energy budgets could alter both the atmospheric circulation and the water cycle (e.g. Gutowski *et al.* 1991; Boer 1993). In addition, this analysis provides a basis for understanding how factors such as atmospheric moisture are related to the surface energy budget.

Table 2 shows the regional annual averages for the surface energy budget from the control experiment and the difference between the deforestation and control experiments over the three study regions (after Zhang *et al.* 1996a). The surface net solar radiation flux has been separated into incident solar radiation at the surface and the reflected solar radiation in order to distinguish between the effect of the distribution and nature of cloudiness and the effect of increased surface albedo in the deforestation simulation. Also the net outgoing longwave radiation is split into upward and downward components in order to distinguish changes in the thermal radiation emitted by the surface and changes in the downward radiation from the atmosphere to the land surface. The column energy budget from the control experiment shows that these two longwave fluxes are the two largest fluxes in the surface energy balance, a fact also stressed by Gutowski *et al.* (1991). In the Amazon, 88% of the incident solar radiation is absorbed by the land surface. The net radiative energy is almost balanced by the loss of surface latent heat and sensible heat. About 64% of the net radiative energy is allocated to the latent heat flux through evapotranspiration and 36% is lost as sensible heat.

Changes in the surface energy budget following deforestation include alterations in reflected solar radiation, the net radiation energy and evaporation (Table 2). The increase in the reflected solar radiation represents the effect of the increased surface albedo. However, a large part of this effect is the result of a rise in incident solar radiation attributable to the decreased cloudiness (Table 2). The net reduction in the solar radiation absorbed by the land surface is 6.7 W m^{-2}. Variations in the longwave radiation, including the increased upward longwave radiation emitted by the land surface and the decreased downward longwave radiation emitted by the atmosphere, make a significant contribution to the changes of surface net radiation. The increase of net outgoing longwave radiation at the surface is larger than the change in the net solar radiation (cf. Gutowski *et al.* 1991). The end result is a net radiative energy loss to the land surface as a result of deforestation.

The decrease in net radiative energy at the land surface is offset by the reduction in latent heat flux and the very small increase in sensible heat flux. It appears that competing changes in surface albedo (regulating the surface net solar radiation) and the changes in surface roughness length (controlling the evapotranspiration), together with the associated alterations in cloud forcing and the longwave radiative processes, combine to produce the small simulated change in surface temperature.

TABLE 2. Annual average of surface energy budget over the Amazon Basin, Southeast
           Asia and tropical Africa

Quantity	Experiment	Amazon	Southeast Asia	tropical Africa
incident solar radiation at the land surface	control	246.9	234.6	259.2
	difference	+10.4	+5.1	+2.7
net solar radiation absorbed at the surface	control	217.9	205.5	220.9
	difference	-6.7	-8.3	-5.6
solar radiation reflected by the land surface	control	29.0	29.1	38.3
	difference	+17.1	+13.3	+8.3
downward longwave radiation at the surface	control	388.5	388.6	382.9
	difference	-6.2	-4.1	-3.7
net upward longwave radiation at the surface	control	61.5	55.9	74.1
	difference	+9.6	+3.7	+4.0
longwave radiation emitted by the land surface	control	450.0	444.5	457.0
	difference	+3.4	-0.4	+0.3
net radiative energy at the surface	control	156.4	149.6	146.8
	difference	-16.3	-11.9	-9.7
latent heat flux	control	99.9	108.1	82.2
	difference	-17.8	-11.1	-6.0
sensible heat flux	control	56.8	41.7	65.0
	difference	+1.6	-0.9	-3.7
net surface energy budget	control	-0.3	-0.2	-0.4
	difference	-0.05	0.04	-0.04
total cloud amount	control	51.0	-	-
	difference	-4.0	-	-

Units are W m^{-2} for all quantities except total cloud (%).

Examining the surface and atmospheric energy budgets has provided several pieces of information useful for understanding the simulated climatic changes after deforestation. The results differ in detail from earlier interpretations. Generally, the cancelling effects of increased surface albedo and decreased evapotranspiration explain the small changes in surface temperature over deforested regions, but the analysis also suggests that:

   i. cloud radiative forcing plays an important role in shortwave radiation
      processes while the influence of increased surface albedo is mitigated
      dramatically by the decrease of cloud amount;

ii. the effect of the change in net longwave radiation is larger than the effect of the shortwave radiation change, a result not emphasised in previous simulations;

iii. changes in cloud radiative forcing increase the daily variability of the surface temperature even though the daily mean surface temperature only shows small changes;

iv. in spite of the increase of surface albedo over the deforested region, the net radiative energy heating the atmosphere is actually increased by the changes in cloud radiative forcing and the longwave radiation;

v. the role of changes in latent heat flux is the most important factor in the net reduction of the atmospheric energy budget.

From this evaluation, it seems reasonable to assert that changes in latent heat flux, themselves the result of the reduction of surface roughness length and the changes in surface radiative energy, are the primary reasons for the stimulated reduction of precipitation over the deforested regions from which rainforest has been removed.

2.4   IMPACTS ON HYDROLOGICAL PROCESSES

There are two ways to look into hydrological processes associated with the deforested areas: the first considers the partition of precipitation over the basin while the second evaluates the sources of precipitation (i.e. advected moisture cf. local sources). Salati and Vose (1984), Salati (1987) and Salati and Nobre (1991) found that roughly 25% of the total precipitation is intercepted by the tropical rainforest and re-evaporated into the atmosphere while around 50% is returned into the atmosphere through the forest transpiration and ground evaporation. Overall, about 75% of the total precipitation is 'captured' by the forest and 'recycled' into the atmosphere.

Lettau et al. (1979) and Brubaker et al. (1993) examined what proportion of precipitation in a continental region is formed from advected (external) moisture delivered by the regional atmospheric circulation, and what proportion from the regional water (internal) recycling by evapotranspiration. In the Amazon Basin, the dominant convergent airflow results in a major part of surface-evaporated water remaining in the basin with only a little being transported out of the basin. This water recycling over a specific region is determined by:

i. how much precipitated water can be returned to the atmosphere as surface transpiration;

ii. to what extent the surface-evaporated water is held in the atmosphere over the same region to become precipitation and to sustain the regional hydrological cycle (e.g. Entekhabi et al. 1992; Brubaker et al. 1993).

Figure 5a shows the hydrological processes represented in the control experiment for the annual averaged case over the whole Amazon Basin: about 41% of the total precipitation is held and recycled into the atmosphere by rainforest transpiration and ground soil evaporation and 24% of the precipitation is intercepted by the tropical

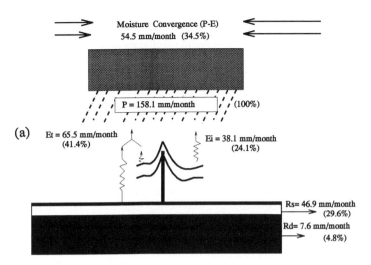

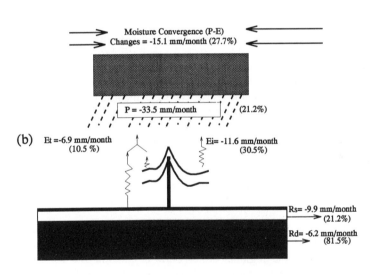

FIGURE 5. Schematic illustration of the hydrological cycle in the control experiment over the Amazon Basin. P is the annually averaged monthly precipitation: Et is the evapotranspiration from the forest and ground, Ei is the rate of precipitation interception by foliage (minus dew drop); Rs is the surface runoff and Rd is the deep ground water runoff. The percentages represent the contribution to the total precipitation; (b) changes in the hydrological cycle over the Amazon Basin after deforestation. Numbers in parentheses are the percentage changes.

forest to be evaporated back into the atmosphere. Consequently, 65% of the total precipitation is held by the rainforest and returned into the atmosphere local to the original rainfall. These percentages are in good agreement with the existing estimations (Salati and Nobre 1991). The remaining 35% of the total precipitation is lost as surface and groundwater runoff which must be balanced by the atmospheric transport (Hartmann 1994).

Replacing the tropical forest with grassland significantly reduces the evapotranspiration and variations in the energy balance can be expected to change the regional atmospheric circulation and external moisture transports. Total precipitation decreases by 21.2% (Figure 5b) and the imposed change in vegetation leads to 30.5% reduction of precipitation interception, which is greater than the reduction of evaporation. Thus, the physiological changes of tropical rainforest, such as the reduction in the leaf area, are of importance in the changes of water recycling over the Amazon Basin. The total basin water recycling reduces by 18.5 mm month^{-1}, and the external moisture source is decreased by 15.1 mm month^{-1} (27.7%). The decrease of moisture convergence contributes to about 45% of the reduction of total precipitation.

This simulation suggests that the reduction of latent heat flux after deforestation is mainly caused by the reduction in interception rather than the decrease of transpiration and ground evaporation. Also the current regional climate and vegetation co-exist in a dynamic system in which the surface vegetation not only controls the regional water recycling but is important in sustaining the favourable dynamic structure of the regional atmospheric circulation as well.

3. Impacts of rainforest removal on regional moisture dynamics

Surface evapotranspiration and sensible heat flux are related to the dynamic structure of the low level atmosphere including the horizontal wind speed, the atmospheric stability and the efficiency of aerodynamic transfer between the low-level atmosphere and the land surface. In a deforestation experiment, one of the major changes imposed is a reduction in surface roughness length, which, as reported in Sud et al. (1996), leads to the decrease of the aerodynamic drag coefficient (CDN) (Dickinson et al. 1993). At the same time, because tropical winds in the boundary layer are thermally driven and frictionally controlled (Sud et al. 1996), the reduction of the surface frictional force should result in stronger surface winds and this may mitigate the effects of the decrease in the drag coefficient.

It is frequently stressed (e.g. Nobre et al. 1991; Henderson-Sellers et al. 1993; McGuffie et al. 1995; Sud et al. 1996) that the alterations of vegetation type can modify the characteristics of the regional atmospheric circulation and the large-scale external moisture fluxes. The changes in the horizontal divergence field can be examined to show variations in the large-scale flow over the region. Deforestation greatly modifies the divergence structure and induces significant positive changes (reduced convergence) at the low-level and significant negative changes (reduced divergence) in the upper levels. In order to quantify how the deforestation throughout the tropics affects moisture transport, the moisture flux across the boundaries of the three deforested regions were examined for the four seasons: DJF, MAM, JJA and SON.

During all four seasons, moisture transport from the Atlantic Ocean into the Amazon Basin is the major moisture source (Figure 6a), in agreement with the analyses of Salati and Vose (1984) and Salati and Nobre (1991). This large moisture source for the Amazon Basin ranges from 3.9×10^8 kg s^{-1} in MAM to 4.75×10^8 kg s^{-1} in SON. Deforestation leads to changes in moisture transport across the eastern boundary in the form of an increase of moisture transport in DJF ($+2.01 \times 10^7$ kg s^{-1}) and a decrease in JJA (-3.44×10^7 kg s^{-1}). Marginal changes occur in MAM and SON but these are not the major reason for the simulated decrease of moisture convergence over the Amazon Basin during the local wet season.

At the western boundary of the Amazon Basin, prevailing easterly trade winds out of the basin result in a loss of water vapour in the lower atmospheric levels which is much weaker than the incoming water vapour flux from the eastern boundary in the control experiment. Deforestation produces large changes here: the reduction of surface friction caused by the imposed deforestation increases the magnitude of the low-level winds and more moisture is conveyed out of the region especially during the seasons of MAM (by about 9.04×10^7 kg s^{-1}) and JJA (2.26×10^8 kg s^{-1}). Compared with the changes at other boundaries, results show that the strengthened outgoing flow at the western boundary of the Amazon Basin is a major reason for the reduction in moisture convergence following deforestation.

Moisture transport crossing the northern and southern boundaries exhibits significant seasonal variability due to the seasonal migration of the Inter-Tropical Convergence Zone (ITCZ)(Figure 6a). In DJF, large amounts of moisture (3.04×10^8 kg s^{-1}) are brought into the Amazon Basin from the northern tropical oceans by the strong ITCZ. This changes dramatically in JJA when the ITCZ is located in the tropical region of the Northern Hemisphere. The northern boundary of the Amazon acts as a sink for moisture in this season, with about 1.21×10^8 kg s^{-1} of humid air being moved out of the basin. The effect of deforestation on the water transport crossing the northern boundary is represented by the decrease of incoming and the increase of outgoing moisture during the four seasons.

Moisture transport crossing the southern boundary of the Amazon Basin could provide an insight into how deforestation can produce climatic changes remote from the location of the surface disturbance. Wang *et al.* (1997) calculated the moisture transport over the region south of the Amazon Basin and reported that the incoming moisture from the Amazon Basin is one of the important components of this region's water vapour budget. In the simulation reported here, deforestation leads to even more moisture being conveyed south from the Amazon Basin in DJF: an increase of 3.96×10^7 kg s^{-1}. This numerical experiment suggests, therefore, that deforestation in the Amazon can affect the precipitation to the south of the basin.

The moisture convergence into the Amazon Basin as a whole is decreased throughout the year following deforestation (Figure 6a). Results confirm that deforestation can reduce the high precipitation regime of the Amazon Basin not only by the reduction of regional water recycling through evapotranspiration from the rainforest but also by the reduction of moisture transport into the area following changes in the regional atmospheric circulation.

It is expected that changes in the atmospheric dynamics will be smaller over the land areas of Southeast Asia than the Amazon Basin because of the smaller deforested area and the strong monsoon circulation. Figure 6b shows the moisture

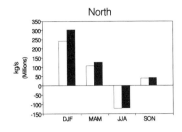

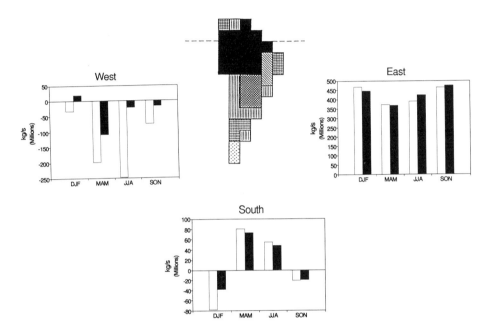

FIGURE 6a. Vertically integrated water vapour transport across the boundaries of the
Amazon Basin for the four seasons. Positive values indicate incoming water
vapour transport.

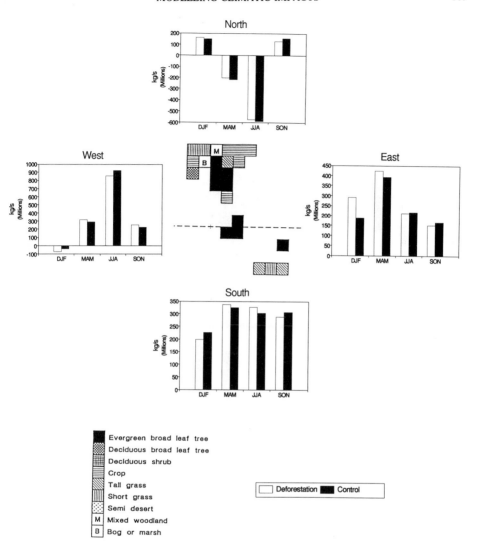

FIGURE 6b. Vertically integrated water vapour transport across the boundaries of
Southeast Asia for the four seasons. Positive values indicate incoming water
vapour transport.

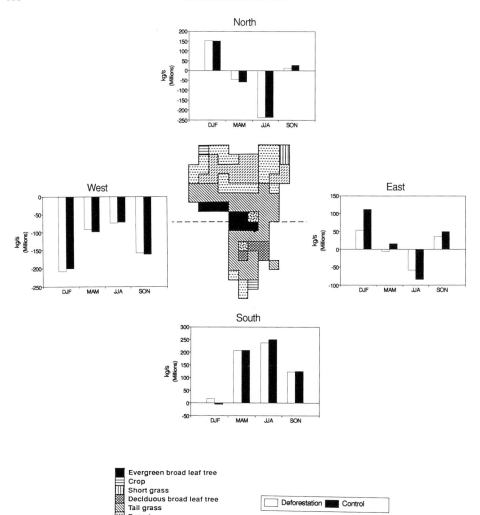

FIGURE 6c. Vertically integrated water vapour transport across the boundaries of
            tropical Africa for the four seasons. Positive values indicate incoming water
            vapour transport.

transport across the boundaries of Southeast Asia (20°S to 20°N, 90°E to 150°E) for the four seasons. In general, moisture from the Indian Ocean is important for the summer monsoon precipitation over Southeast Asia especially over the Indochina peninsula. The west Pacific warm pool is also important, but in this model simulation its contribution is not as significant as that from the Indian Ocean. Large amounts of water vapour are lost across the northern boundary of Southeast Asia, implying that the rainforest region is a significant source of moisture for the rest of Asia. Following deforestation, the moisture flowing out of Southeast Asia which supports the high precipitation during the summer monsoon over East Asia is reduced from 5.87×10^8 kg s^{-1} to 5.74×10^8 kg s^{-1}.

The striking seasonal variation in the Southeast Asian monsoon prompts changes in the hydrological processes during DJF that are quite different from those during JJA. Moisture is lost at the western boundary of Southeast Asia in DJF with large negative values in Figure 6b. In contrast, there are large incoming fluxes of moisture from the west Pacific Ocean crossing the eastern and southern boundaries. Deforestation leads to a reduction in incoming moisture across the southern boundary of Southeast Asia but an increase in the northern and eastern boundaries. The net incoming moisture over Southeast Asia as a whole is actually increased from 5.30×10^8 kg s^{-1} to 5.81×10^8 kg s^{-1} following deforestation which explains the smaller reduction in precipitation in DJF than in JJA over over deforested continent (cf. Zhang *et al.* 1996b). In addition Figure 6b shows that total incoming moisture is increased during MAM and decreased in SON following deforestation.

There are significant seasonal variations in the incoming moisture transport over tropical Africa. In the seasons of MAM and SON, the ITCZ, located near the equator, transports a huge amount of moisture from the Atlantic Ocean into this region at the southern boundary and results in large moisture convergence, e.g. about 6.87×10^7 kg s^{-1} in MAM; while, in JJA and DJF, the location of the ITCZ is distant from the forest region and less moisture is conveyed into it. Indeed, in JJA, there is even a water vapour loss of -1.42×10^8 kg s^{-1}. This explains the double peak in the seasonal precipitation over the tropical rainforest region over Africa seen in Figure 3g. Following deforestation, moisture transport over tropical Africa is decreased from 5.81×10^7 kg s^{-1} to 1.47×10^7 kg s^{-1} in DJF and from 4.41×10^7 kg s^{-1} to 1.46×10^7 kg s^{-1} in SON. However, there are only very small changes in MAM and JJA. The incoming moisture in MAM is reduced from 6.87×10^7 kg s^{-1} to 6.43×10^7 kg s^{-1} and the moisture loss is reduced from -1.42×10^8 kg s^{-1} to -1.34×10^8 kg s^{-1}. The small changes in these seasons could be used to explain the small changes in precipitation in JJA and even some increases in MAM following deforestation (cf. Zhang *et al.* 1996b). The impacts of the imposed deforestation detected in the regional changes in moisture flux have the potential to induce change in the large-scale atmospheric flow and possibly global scale impacts. This possibility and the underlying mechanisms are discussed in the next section.

4.   Impacts of tropical deforestation on the global climate

While local and regional impacts of tropical deforestation have been identified in GCM simulations in recent years, one challenging question left unanswered is whether there are impacts of tropical deforestation on the global climate. The issue of distant-from-deforestation effects becomes more important when it is recognised that international co-operation is required in order to try to slow deforestation in tropical regions.

The first GCM simulation used to try to estimate the impacts of deforestation on the global climate was that of Henderson-Sellers and Gornitz (1984). These authors suggested that, based upon their model's simulation, it was difficult to assert that tropical deforestation could produce global scale influences. However, using a more sophisticated land-surface scheme and a longer integration, McGuffie *et al.* (1995) reported that tropical deforestation could result in global scale climatic disturbances. Sud *et al.* (1996), employing a quite different GCM and a different land-surface scheme, also reached a similar conclusion but the means of producing the distant climatic impacts remained uncertain.

The circulation of the atmosphere over the tropics is characterized by the meridional Hadley Circulation and the zonal Walker Circulation (Figure 2; Oort and Peixoto 1983). Namaguti (1993) defined an important role for evapotranspiration in the dynamics of the Hadley circulation; his results indicating that a small fractional change in the evapotranspiration causes a large change in the net energy supply and results in a large modification to the meridional structure of the Hadley circulation. Therefore, it might be anticipated that deforestation, which has been shown to dramatically affect the regional evapotranspiration over the disturbed regions, might have considerable impacts on the regional and global circulations. In this section, the possible impacts of tropical deforestation on the global climate are considered. The dynamical mechanisms possibly responsible for changes distant from the deforested regions are explored by analysing the importance of tropical rainforests in sustaining the Hadley and Walker circulations. A possible explanation for the simulated teleconnections is developed based upon Rossby wave propagation.

Since the tropics are the primary energy source (high solar radiation) and the moisture source (high sea surface temperatures) for the global climate, changes over the tropical regions might influence the global climate in a manner similar to the propagation of El Nino-Southern Oscillation (ENSO) disturbances into the extra-tropics (e.g. Webster 1982; Ting and Hoerling 1993; Zhang *et al.* 1996b). The dynamical processes transporting the regional influences outside the deforested regions are the changes in the tropical Hadley Circulation and Walker Circulation, and Rossby wave propagation excited by the effects of deforestation.

Following deforestation, changes in the meridional streamfunction indicate that the strongly ascending branch of the Hadley cell is considerably weakened due to the net loss of energy for the regional atmospheric circulation (Zhang *et al.* 1996b). Meanwhile, the large northward streamfunction gradient over the subtropical region of the Northern Hemisphere (from 10°N to 30°N) demonstrates that the descending flow over the subtropics of the Northern Hemisphere is reduced following deforestation. The ascending branch over the middle latitudes of the Northern Hemisphere in January is also weakened. Similarly large changes in the Hadley Circulation are seen in July. These include the significant reduction of upward motion in the Northern Hemisphere ascending branch (around 10°N) and the weakening descending branch in the Southern Hemisphere (10°S). Large changes also occur in the high latitudes.

Four ascending branches of the Walker Circulation are clearly represented over Southeast Asia and the west Pacific Ocean, tropical Africa and tropical South America co-located with the tropical rainforests. Following deforestation, the changes in the Walker Circulation reveal the linkages in the tropical circulation. Firstly, the ascending branches located over the tropical rainforest are significantly weakened over tropical South America (80°W to 60°W), tropical Africa (0 to 30°E) and Southeast Asia (110°E to 150°E). In addition, changes are also stimulated over the tropical oceans. The strong, ascending branch over continental Southeast Asia is moved eastward into the west and central Pacific Ocean as reported in McGuffie *et al.* (1995). The circulation over the Indian Ocean is also modified by the enhanced ascending and descending motions over the eastern and western Indian Ocean, respectively. The influence of deforestation on the Walker circulation over the tropical oceans provides a dynamical explanation for the simulated changes in the sea surface temperature found by McGuffie *et al.* (1995).

Rossby wave propagation has been proposed as a possible mechanism whereby the simulated disturbances of the local scale evaporative fluxes can penetrate to higher levels in the atmosphere and propagate to higher latitudes (particularly in the winter hemisphere). The analysis of Zhang *et al.* (1996a,b) indicates that tropical deforestation, which reduces the surface evaporation and diminishes the release of condensation heat by affecting tropical convective precipitation, may excite large-scale Rossby waves which propagate towards the middle and high latitudes in both hemispheres and result in the global impacts of tropical deforestation.

Overall, changes in the tropical Hadley and Walker circulations are identified as an important consequence of deforestation. Changes in the Hadley Circulation can induce disturbances in meridional energy and moisture transports from the tropical region to the middle and high latitudes in both hemispheres. Meanwhile, the Walker Circulation is affected by the deforestation not only over the deforested regions but also over the tropical oceans. Both the changes in the Hadley Circulation and in the Walker Circulation provide some explanations for the variations in the general circulation simulated over extra-tropical regions. Analysis of geopotential height and temperature alterations at high levels in the model simulation suggest the possibility of the propagation of Rossby waves forced by the land-surface disturbance. These results indicate that the impacts of tropical deforestation can produce extra-tropical climatic signals similar in nature to those caused by ENSO events.

5.   Summary: the climatic impacts of rainforest destruction

Tropical deforestation greatly reduces biodiversity and produces an additional source of atmospheric CO_2 and other trace gases. Forest burning also forms aerosols. In addition to these massive changes, the act of deforestation modifies the surface state to such an extent that climatic changes are predicted to occur, locally, regionally and perhaps even globally.

The climatic impacts stimulated in the Amazon Basin, Southeast Asia and tropical Africa are regionally specific. Significant climatic changes are stimulated in the Amazon Basin but the variations in Southeast Asia and tropical Africa are smaller, especially those in local precipitation. There is an interaction between the changes in surface evapotranspiration and the changes in total precipitation. Following deforestation, the reduction in precipitation is initially due to the decrease in surface evapotranspiration (because the net radiative energy in the atmosphere is increased over all three regions), and this reduction in the total precipitation further reduces the evapotranspiration. At the same time, the decline in the latent heat flux, because of the weakened surface evapotranspiration, leads to a net reduction in the atmospheric energy budget. Thus the regional atmospheric circulation is weakened and less water vapour is delivered into the deforested regions.

The rather small changes in surface temperature detected in these simulations are the result of the compensating effects of the reduction in the surface net radiation and the reduction in the surface evapotranspiration. Changes in sensible heat flux are generally not of importance in determining the surface energy budget, except in tropical Africa: statistical analysis indicates that variations in the sensible heat flux are controlled by changes of surface net radiative energy and alterations in surface evapotranspiration (Zhang *et al.* 1996a).

The results presented here clearly show a sensitivity of the local climate to the removal of tropical forest. The local recycling of moisture by the vegetation canopy is confirmed to be a vital component in understanding the disturbances to the local hydrology. Moreover, the scale of moisture convergence changes and possibly also cloud and convection changes, is such that there is a possibility that non-local climatic impacts may also occur.

Calculations of the water vapour transports over the three deforested regions demonstrate that a larger reduction in the total precipitation than in the surface evapotranspiration is the result of decreased moisture convergence following deforestation. There is an interactive process between the variations in surface evapotranspiration and the changes in the total precipitation. In the deforested conditions, the reduction in precipitation is partly caused by the reduction in the surface evapotranspiration (because the net radiative energy in the atmosphere is increased over all three regions), while the alterations in the regional and large-scale atmospheric circulation lead to a further reduction of regional precipitation, which can further reduce the evaporation. Consequently, less water vapour is delivered into the deforested regions except for the special case over tropical Africa in March. Over the Amazon Basin, the smooth land surface following deforestation enlarges the outflow across the western boundary of the basin. This change brings more water vapour out of the basin and is underlined in this study to explain the reduction of water vapour convergence over the Amazon.

Modification of the model surface parameters to simulate tropical deforestation produces significant modifications of both the Hadley and Walker circulations, which result in changes distant from the region of deforestation. The mechanism for propagation to middle and high latitudes of disturbances arising from tropical deforestation is based on Rossby wave propagation. These propagating waves, which have, in the past, been associated with the extra-tropical influences of ENSO events (e.g. Webster 1982; Ting and Hoerling 1993), provide a pathway for the dispersion of the tropical disturbances to high latitudes.

It must be noted that the analysis presented here has considered *only* the potential effects on climate of the physical changes caused by tropical forest removal. In addition to these, the impacts of increased greenhouse gases, atmospheric aerosols and altered tropospheric and stratospheric chemistry must be considered. It seems clear that numerical simulations of climatic impacts of tropical deforestation add further reasons for greatly reducing, or ending totally, rainforest destruction.

References

Bastable, H.G., Shuttleworth, W.J., Dallarosa, R.L.G., Fisch, G. and Nobre, C.A. (1993). Observations of climate, albedo, and surface radiation over cleared and undisturbed Amazonia forest, *International Journal of Climatology*, **13**, 783-796.

Boer, G.J. (1993). Climate change and the regulation of the surface moisture energy budgets, *Climate Dynamics*, **8**, 225-239.

Brubaker, K.L., Entekhabi, D. and Eagleson, P.S. (1993). Estimation of continental precipitation recycling, *Journal of Climate*, **6**, 1077-1089.

Charney, J.G. (1975). Dynamics of deserts and drought in the Sahel, *Quarterly Journal of the Royal Meteorological Society*, **101**, 193-202.

Cunnington, W.M. and Rowntree, P.R. (1986). Simulation of the Saharan atmospheric dependence on moisture and albedo, *Quarterly Journal of the Royal Meteorological Society*, **112**, 971-999.

Dickinson, R.E. and Henderson-Sellers, A. (1988). Modelling tropical deforestation: a study of GCM land-surface parameterizations, *Quarterly Journal of the Royal Meteorological Society*, **114**, 439-462.

Dickinson, R.E. and Kennedy, P. (1992). Impacts on regional climate of Amazon deforestation, *Geophysics Research Letters*, **19**, 1947-1950.

Dickinson, R.E., Henderson-Sellers, A., Kennedy, P.J. and Giorgi, F. (1993). Biosphere-Atmosphere Transfer Scheme (BATS) Version 1e as coupled to the NCAR Community Climate Model, NCAR Technical Note NCAR/TN-387+STR, National Center for Atmospheric Research, Boulder, Colorado.

Dickinson, R.E., Henderson-Sellers, A., Kennedy, P.J. and Wilson, M.F. (1986). Biosphere-Atmosphere Transfer Scheme (BATS) for the NCAR Community Climate Model. NCAR Technical Note NCAR/TN-275+STR, National Center for Atmospheric Research, Boulder, Colorado.

Dirmeyer, P.A. (1992). GCM studies of the influence of vegetation on the general circulation: the role of albedo in modulating climate change, Unpublished PhD thesis, Department of Meteorology, University of Maryland.

Entekhabi, D., Rodriguez-Iturbe, I. and Bras, R. (1992). Variability in large-scale water balance with land surface interaction, *Journal of Climate*, **5**, 798-813.

Figueroa, S. and Nobre, C. (1990). Precipitation distribution over central and western tropical South America, *Climanalise*, **5**, 36-45.

Giambelluca, T.W. (1996). Tropical land cover change: characterizing the post-forest land surface, in T.W. Giambelluca and A. Henderson-Sellers (eds.), *Climatic Change: developing southern hemisphere perspectives*, Wiley, Chichester, pp. 293-318.

Gutowski, W.J., Gutzler, D.S. and Wang, W-C. (1991). Surface energy balance of three general circulation models: implications for simulating regional climate changes, *Journal of Climate*, **4**, 121-134.

Hartmann, D.L. (1994). *Global Physical Climatology*, Academic Press, New York.

Henderson-Sellers, A. and Gornitz, V. (1984). Possible climatic impacts of land cover transformations, with particular emphasis on tropical deforestation, *Climate Change*, **6**, 231-235.

Henderson-Sellers, A., Durbidge, T.B., Pitman, A.J., Dickinson, R.E., Kennedy, P.J. and McGuffie, K. (1993). Tropical deforestation: modelling local to regional-scale climatic change, *Journal of Geophysical Research*, **98**, 7289-7315.

Hendon, H.H. and Woodberry, K. (1993). The diurnal cycle of tropical convection, *Journal of Geophysics Research*, **98**, 16623-16637.

Lean, J. and Rowntree, P.R. (1992). A GCM simulation of the impacts of Amazonian deforestation on climate using an improved canopy representation, Climate Research Technical Note No. 26, Hadley Centre for Climate Prediction and Research, Meteorological Office, Bracknell, Berkshire, U.K.

Lean, J. and Warrilow, D.A. (1989). Simulation of the regional climatic impact of Amazon deforestation, *Nature*, **342**, 411-413.

Lettau, H., Lettau, K. and Molion, L.C.B. (1979). Amazonia's hydrologic cycle and the role of atmospheric recycling in assessing deforestation effects, *Monthly Weather Review*, **107**, 227-238.

McGuffie, K., Henderson-Sellers, A., Zhang, H., Durbidge, T.B. and Pitman, A.J. (1995). Global climate sensitivity to tropical deforestation, *Global Planetary Change*, **10**, 97-128.

Mintz, Y. (1982). The sensitivity of numerically simulated climates to land surface conditions, in P.S. Eagleson (ed.), *Land Surface Processes in Atmospheric General Circulation Models*, Cambridge University Press, Cambridge, pp. 109-111.

Mylne, M.F. and Rowntree, P.R. (1992). Modelling the effects of albedo change associated with tropical deforestation, *Climatic Change*, **21**, 317-343.

Nobre, C.A., Sellers, P.J. and Shukla, J. (1991). Amazonian deforestation and regional climatic change, *Journal of Climate*, **4**, 957-988.

Namaguti, A. (1993). Dynamics and energy balance of the Hadley circulation and the tropical precipitation zones: significance of the distribution of evaporation, *Journal of Atmospheric Science*, **50**, 1874-1887.

Oort, A.H. and Peixoto, J. (1983). Global angular momentum and energy balance requirements from observations, *Advances in Geophysics*, **25**, 355-490.

Polcher, J. and Laval, K. (1994a). The impact of African and Amazonian deforestation on tropical climate, *Journal of Hydrology*, **155**, 389-405.

Polcher, J. and Laval, K. (1994b). A statistical study of the regional impacts of deforestation on climate in the LMD GCM, *Climate Dynamics*, **10**, 205-219.

Randall, D.A., Harshvardhan, and Dazlich, D.A. (1991). Diurnal variability of the hydrologic cycle in a general circulation model, *Journal of Atmospheric Science*, **48**, 40-62.

Salati, E. (1987). The forest and the hydrological cycle, in R.E. Dickinson (ed.), *The Geophysiology of Amazonia*, Wiley, New York, pp. 273-296.

Salati, E. and Vose, P.B. (1984). Amazon Basin: a system in equilibrium, *Science*, **225**, 129-137.

Salati, E. and Nobre, C.A. (1991). Possible climatic impacts of tropical deforestation, *Climatic Change*, **19**, 177-196.

Sud, Y.C., Walker, G.K., Kim, J-H., Liston, G.E., Sellers, P.J. and Lau, K-M. (1996). Biogeophysical effects of a tropical deforestation scenario: a GCM simulation study, *Journal of Climate*, **9**, 3225-3247.

Ting, M. and Hoerling, M.P. (1993). Dynamics of stationary wave anomalous during the 1986/87 El Nino, *Climate Dynamics*, **9**, 147-164.

Wang, M., Paegle, J. and Paegle, J.N. (1997). Water vapour balance over North and South America, *Journal of Climate* (accepted).

Webster, P.J. (1982). Seasonality in the local and remote atmospheric response to sea surface temperature anomalies, *Journal of Atmospheric Science*, **39**, 41-52.

Williamson, D.L., Kiehl, J.T., Ramanathan, V., Dickinson, R.E. and Hack, J.J. (1987). Description of the NCAR Community Climate Model (CCM1), NCAR Technical Note, NCAR/TN-285+STR.

Zhang, H., Henderson-Sellers, A. and McGuffie, K. (1996a). Impacts of tropical deforestation I: process analysis of local climatic change, *Journal of Climate*, **9**, 1497-1517.

Zhang, H., McGuffie, K. and Henderson-Sellers, A. (1996b). Impacts of tropical deforestation II: the role of large-scale dynamics, *Journal of Climate*, **9**, 2498-2521.

Zhang, H., Henderson-Sellers, A., McAvaney, B. and Pitman, A. (1997). Uncertainties in GCM evaluations of tropical deforestation: a comparison of two model simulations, in W. Howe and A. Henderson-Sellers 9eds.), *Assesing Climate Change: results from the Model Evaluation Consortium for Climate Assessment,* Gordon and Breach, Sydney, Australia, pp. 323-355.

Dr. Kenneth McGuffie, Department of Applied Physics, University of Technology, Sydney, New South Wales 2007, Australia;

Prof. Ann Henderson-Sellers, Royal Melbourne Institute of Technology, Melbourne, Victoria 3001, Australia;

Dr. Huqiang Zhang, Bureau of Meteorology Research Centre, P.O. Box 1289K, Melbourne, Victoria 3001, Australia

10. CONCLUSION

B.K. Maloney

1. Introduction

It would be easy to conclude a series of essays by summarising the ideas and arguments put forward by each author chapter by chapter without attempting to draw the various strands together. It is certainly a way of avoiding potential conflict between the editor and the individual contributors, as no editor can expect to agree with everything that has been written by each individual author, nor can that editor expect total agreement with what he has said, or, indeed, the way in which he has edited. It is, therefore, necessary to state at the outset that what follows is a personal and not necessarily a collective view.

This book has attempted to explore the past, present, and possible future of the tropical rainforest. To examine each one of these topics throughly would require a whole book devoted to every individual topic, such is the bulk of the literature, if not the knowledge, of the rainforest. It is accepted that limitations have been imposed in the consideration of the rainforest presented here because a wide range of topics could not be examined relevant to the present and future of the rainforest especially. It would be interesting to tackle the problem including inputs from a wider range of specialists on a regional basis and then try to integrate the findings in an overview. This would take a great deal of time and patience in assembling and editing contributions and by the time the book was published both the individual reports and the overview might be quite outdated. As a result of these logistical considerations, the constraints of space, the aim to reach a multi-disciplinary audience, but without insulting the intelligence of that audience by producing superficial accounts of each topic covered, and the aim to be non-partisan, apolitical and academically careful rather than histrionic in approach, and, of course, to publish quickly, this book has many gaps in cover and some may argue imbalances between chapter contents. However, the authors hope that the approach, warts and all, is novel and stimulating and that will encourage people in future not to become lost in the wood of their own discrete specialisms for the increasing lack of trees. The main message is that an integrated view of the past and present of the tropical rainforest is necessary to examine its future. Obviously because of the limitations outlined above it has not been possible to produce as holistic an account as could be desired.

2. The physical background

As Flenley (1979) pointed out, for a long time people regarded the tropical rainforest and its environment as stable, unchanging over time, until people came along and started to destroy it. This is not so, climate has changed over long and short term cycles (Maley 1996; Thomas, this volume) and this has affected, and continues to affect both geomorphological and soil evolution. Changes have been particularly notable in areas outside the stable shields, mainly located in Africa and South America, but even there the rainforest has expanded and contracted over time. In Southeast Asia, often referred to as a 'maritime continent' because the Sunda-Sahul Shelf is so shallow and must have been exposed several times during the glacial-interglacial cycles of the last 2 m years, stability over any length of time cannot have been the norm especially in the mountains

195

B.K. Maloney (ed.), Human Activities and the Tropical Rainforest, 195-206.
© *1998 Kluwer Academic Publishers. Printed in the Netherlands.*

and tectonically active areas (cf. Garwood *et al.* 1979; Johns 1986, 1988). Indeed *c.* 74,000 years ago this area witnessed the most massive volcanic eruption of comparatively recent times (Chesner *et al.* 1991) and there had been several other earlier, less well documented, eruptions. Some (Dawson 1992) suggest that the Late Quaternary eruption may have triggered the last ice age. Whether or not it left the vegetation of north Sumatra floristically depauperate, or triggered the rapid speciation of the Dipterocarpaceae, the main tree family of the lowlands, remains to be demonstrated, but Whitten *et al.* (1984) argued that it left it faunistically the poorer. It is wise to view what is happening to the rainforest today within that perspective, particularly as far as conservation of what remains is concerned. Volcanic activity and burning are associated, but natural burning can occur even in lowland humid tropical rainforest, and, indeed, on peat swamps, where it can be the result of lightening strikes (Anderson 1964, 1966; Brunig 1964). Even normally everwet areas can experience periods of drought (Baillie 1976; Brunig 1971) when fire can spread rapidly, as it did in Kalimantan during 1982-83 (Beaman *et al.* 1985; Malingreau *et al.* 1985; Goldammer and Seibert 1990) and Late Quaternary and Holocene radiocarbon dates have been obtained (Goldammer and Seibert 1989) from lowland charcoal in Borneo. Fire is a normal event in the areas of more seasonal forest (cf. Stott 1985, 1986, 1988, 1990). However, it should be said that there is much dispute as to whether or not fire in humid tropical areas is of natural or human origin. This can be typified by the debate which continues about the possibility of early Aboriginal burning particularly in the rainforest areas of northeast Australia (cf Anderson 1994; Clark 1983; Kershaw 1986, 1994,1995). The case has not, so far, been satisfactorily proven either way.

It can be argued, as Thomas (this volume) has done, that soil erosion as a result of forest removal on slopes, by natural processes or due to human impact, is not of itself a bad thing in terms of a geological perspective of landscape evolution as it uncovers unweathered, or only partly weathered, rock to the processes of natural destruction which release plant available nutrients to the soil, and the movement of the soil downslope or into the valleys can be regenerative of the soils of those areas, depending upon the chemical and physical nature of the material being moved.

In the early days of European exploration it was thought that all rainforest tropical soils must be very fertile because of the luxuriant nature of the vegetation it supported, and numerous disasters occurred as a result when attempts were made to spread European styles of agriculture to the tropics. In time it came to be recognised that gardening makes sound ecological sense, especially on sloping land, but even the lowland planters began to inter-plant to keep the soil surface covered and protected from the baking sun and the intense rainfall. Then the complex nature of nutrient cycling in rainforests was rediscovered. It had been appreciated much earlier by Alexander von Humboldt, and has been resurrected more recently as the 'web of life' concept in plant ecology. Humboldt has been described as the last scientist who could seriously think of providing an overview of the whole natural world (Bowler 1992: 208) but nobody took up the task of completing the universal survey which he began in the tropical Americas, although his influence was pervasive during the rest of the 19th century, because he founded no coherent school of thought. The 'web of life' idea came back into the biological sciences through the physical sciences with the emergence of general systems theory. Within that context, both the 'web of life' idea and general systems theory stresses the intricate inter-relationships between the inanimate and living world.

In time it became orthodox to regard all tropical soils as well weathered and poor in plant nutrients, with the main nutrient store in the vegetation itself (cf Longman and Jenik 1987). A view which is the product of a lack of detailed knowledge of the soils, landforms and rainforest of specific areas. Forest clearance came to be regarded as unwise, not only because of the destruction of vegetation, but because the expected soil fertility in terms of agricultural or horticultural productivity did not merit it as the main source of plant nutrients, the plants themselves, are destroyed. Now, as Nortcliff in Chapter 3 indicates, it is increasingly recognised that the nature of tropical soils is as varied as it is elsewhere, that origins and distributions are complex, and that quite chemically rich soils derive from some volcanic rocks, such as basalt, and, of course, from limestone. There is much merit in the idea that it is not sensible to fell forest with underlying poor soils as, within a short period of time, soil fertility is exhausted and more forest has to be cleared to maintain production. It is common sense to clear land where the soils are chemically and physically good, and to intensify production on these, but we still do not have detailed information for many of the rainforest regions concerning where such soils are located and social, political and long and short term economic targets over-ride the concept of making haste slowly. The result is that geomorphic and soil factors are not given perhaps the same consideration that swiddeners might give them when forest is cleared for commercial agriculture, while loggers are unlikely to pay much attention to the physical environment at all except in terms of how accessible it is, as this will affect profit margins.

As Joseph Conrad said, not in one of his tropical books, but somewhere, if the editor remembers correctly, in 'The Secret Agent', "life does not bear looking in to"; that reality is complex. The principle of Occam's razor is only useful as a teaching tool, not as a means of explaining how the soils and geomorphology of the rainforest areas have evolved in the past, are changing at the present, and would change in the future regardless of increased human impact. It does not need the attention of climatic modellers to tell us what the future of the uplands is likely to be without reafforestation of some nature, even if it is monocultural. That is of increased soil erosion, with consequences for settlements downslope and downvalley: less retention of water by the soil and vegetation, of quicker release water to the streams and rivers, and consequent flooding in the lowlands, silting up of dams, etc. So, more consideration of the physical factors of the environment is needed for rational planning of land use, especially in upland rainforest areas. The problems touched on in this book are not unique to the tropics but occur wherever there is deforestation of uplands. Most tropical rainforest exists, or once existed, on high plateaux or in the lowlands and much attention has been given to the effects on soils of removing forest from such areas. The effects on changing hydrological patterns and on landscape evolution has received less study than perhaps it should have done, perhaps because environmental change is perceived as less dynamic, and therefore less interesting to investigate, from an academic point of view anyway, than the uplands and coastal margins, but studies are now being made (cf. Malmar 1992).

3.   The human past of the tropical rainforest

While it is difficult to generalise about the physical past of the tropical rainforest, it is even more difficult to generalise about its human past. We cannot construct good location maps for sites at different times due to dating problems or construct palaeopopulation maps. It is rare to find evidence for some contiguity of occupance in single areas over even a few thousand years but it sometimes occurs, e.g. at Niah Cave, Sarawak (Harrison 1996) and people have lived in the rainforest or rainforest transition areas for a long time, e.g. in Africa (Clark and van Zinderen Bakker 1962, 1964). However, there is a particular paucity of archaeological sites predating the Holocene and those which have been discovered mainly seem not to be from areas of humid tropical rainforest. It is not surprising therefore that late occupance of the rainforest has become the orthodox idea although there are occasional, and usually controversial, discoveries such as those from Brazil (Meltzer *et al.* 1994; Guidon *et al.* 1996; Parenti *et al.* 1996), which suggest early occupation by hunter-gatherers. However, such finds are not early in terms of the span of human evolution, but long predate any evidence for agricultural use of the land.

The earliest well documented evidence for agriculture and animal domestication comes not from the humid tropics but from the Near East and predates the generally accepted date given for the opening of the Holocene of 10,000 B.P. (Roberts 1989) but agriculture was soon invented, apparently independently, in several parts of the world, for reasons which cannot be clarified but only speculated about. Presently speculation seems to be concentrated upon the climatic change hypothesis. It is assumed, that like the Maya of later times (cf. Gomez-Pompa 1987), the occupants of the rainforest became aware of the value of certain plants to their well being and became agents for their dispersal and survival. However, root crops, which were later taken in to cultivation, have been used for a long time (cf. Loy *et al.* 1992) although remains are seldom found preserved, but there are exceptions (cf. Hather and Hammond 1994). There is some slight indication in the fossil record of *Canarium* from Late Quaternary and early Holocene contexts in Melanesia (cf. Maloney 1996) that people may have begun to stimulate genetic alterations as a result of their conservation efforts, but the degree of genetic change required to produce large fruited varieties of a wild tree might not be very large, perhaps only a single mutation. It may be that it was only when plants were selectively planted on a larger scale or when it was plants that were genetically plastic were used that major changes began to take place.

None of the cereals originated in the tropical rainforest, but, of course, important root crops like yams did. The lingustic evidence from ancient Bantu suggests a possible 5000 year use of the yam and oil palm in West Africa (Ilife 1995). So the major evolutionary effect that humans have had on widely used crop plants does not seem to have begun within the rainforest, and it is likely that it was not a product of the rainforests. However, when we begin to look at areas within the rainforest region, but above the altitude of rainforest *sensu stricto*, above about 1000 m altitude, a different pattern of human usage begins to emerge although the information, again, is both fragmentary and site specific, and the dangers of adopting generalisations based on single occurrences needs to be re-emphasised. The Kuk site in the highlands of Papua-New Guinea is remarkable (Bayliss-Smith and Golson 1992), but there is some evidence that similar sites exist on the island, although they have not been excavated. Kuk is exceptional for many reasons: it has evidence for independent development of drainage techniques (Gorecki 1986), perhaps using only digging sticks, but it is not possible to be certain what was grown. There is some suggestion that banana phytoliths from a cultivated species are present (Wilson 1985) but no finds of root crops, such as provide the staple in the area today occur. It is supposed that some form of shifting cultivation on dry land was practised before the swamps were used and there are pollen records from various parts of the tropical mountains which indicate that swiddening has a fairly lengthy prehistory, but, as with evidence for occupance, finds of habitations, the record is sketchy. With Kuk, however, the suggestion is that this occurred at or very near the opening of the Holocene.

Information about the advent of settled agriculture is also late, and in no instance can it be proved to be entirely based on a root crop subsistence. The earliest cereal which seems to have been grown in Central and South America is maize (Bush *et al.* 1989), and in Asia rice (Glover and Higham 1996), but the latter was almost certainly an introduction from non-tropical areas where the wild species have their natural home. However, the piecemeal information from the tropics generally suggests that agricultural intensification began around 2500-2000 B.P., but we do not know why. It may relate to increasing population pressure on land, be a result of climatic change, or a combination of both. As far as the pollen analyst is concerned, the fossil evidence for a change to drier conditions is the same as that for the introduction of agriculture, except where the pollen of plants which were probably cultivated, such as maize, is found or where there is plant macrofossil data from properly dated archaeological contexts. Much of the rainforest human past is a mystery and will, doubtless, always remain so.

4.   The human present of the rainforest

It is very difficult to cogently summarise the human present of the rainforest because the present for the tribal peoples is so different from that for the miners, the loggers, the tree planters, and the cattle grazers. It is even more different for the village, town, and city dwellers who now occupy land on which rainforest once stood, for the urban resident of Belo Horizonte and the Amazonian Yanomami tribesman. One could say, but the rainforest is not polluted. This is not always the case: mining leads to pollution e.g. by mercury, used to refine gold, which poisons the rivers in parts of Amazonia and elsewhere, while burning by swiddeners releases solids, ash, and invisable pollutants, CO_2, into the atmosphere. Oil extraction also pollutes. Kane (1997) has shown the horrifying effects that this has had on the environment and the lives of the Huaorani and other forest peoples of Ecuador. It was partly because of the varied nature of human activity in the rainforest today that it was decided to select individual topics for more detailed attention rather than to attempt to produce an overview which might result in meaningless generalisations being made. In Chapter 5 Ellen considered the relationship of the indigenous peoples to the forest, and in Chapter 6 Furley the evolution of land-use in Central America from Mayan times, Fraser (Chapter 7) the evolution of land-use in Indonesia from pre-Dutch colonial times, colonial times, to the present and, finally, Barrow examined the spread of *Eucalyptus* monoculture to replace tropical rainforest in Chapter 8. This latter theme could have also been examined for an individual region, but is one better studied from a global perspective as it illustrates how diversity it being replaced by monotony as the rainforest countries become increasingly enmeshed in the global economy. Like the rainforest itself, the content is therefore, heterogenous, but it could be argued that this is a strength rather than weakness of the book. It illustrates that there are a diversity of ways in which the tropical rainforest and its aftermath can be considered, and it is impossible to cover all aspects of it.

In most instances sustained occupation of the lowlands appears to have been later than the highlands and there are generally no strong signs of major technological innovation in the landscape from the earliest agricultural contexts except from Central America. Presently, swiddening, shifting cultivation, is still more a characteristic of the higher land, but this may be because recent evidence of lowland swiddening has been less well documented. The lowland pollen records are too few in number to argue that swiddening was a widespread lowland practice which was pushed to the hills with those who practised it by later invasions of sedentary agriculturalists.

By the time Europeans arrived swiddening was long established in highland areas and at low population densities was probably an effective usage for sloping land, although there is much controversy about the environmental benefits and disadvantages (de Selincourt 1996) of this form of land use apart from the obvious one that the higher the population density, the more frequent clearance is likely to be and the less chance the forest has to recover. The impact of the farmers on the indigenous peoples of the forest in pre-European times is difficult to establish although there are indications that in Indonesia at least of trading contacts from early times linked to the relationships between the Arabs, Indians, and Chinese (Reid 1995). While it is difficult to discern if shifting cultivators ever displaced the people of the forest proper before recent times of land hunger, as indicated by Ellen, the forest people have a special relationship with their environment, a relationship which must differ from that of the swiddener, and even more so from that of the commercial logger and the forester. Not only do they have a great deal of knowledge about the usefulness to themselves of certain plants and animals but their mythology revolves about it, so does their concept of time and space, whereas one might expect the swiddeners, and perhaps, to some extent, the foresters, concept of time to relate to the cycle of crop planting and harvesting, how much energy is needed to fell the forest and how big an area needs to be felled to provide an adequate food supply. In most instances they forage the forest in addition to cultivating the land, but their relationship with the forest is more distant as they are prepared to destroy it. Both groups cannot be readily separated out, as there are overlaps, but the basic survival strategy, and, therefore, concept of and relationship with the forest differs. The commercial loggers are entirely different animals, sometimes concerned only with extraction of individual species, what perhaps could be called monoculling, neglecting native rights over land (Primack 1991) and at most interested only in the timber trees, not the herbs, the shrubs, the epiphytes and the smaller or sapling trees but their long term impact varies (Borhan et al. 1990; Brooks et al. 1993; Cannon et al. 1994; Woods 1989) with the methods and intensities of harvesting used.

The relationship of the swiddener to the forest requires further investigation from a longer term and more recent perspective. Swiddening is an extremely varied form of agriculture which can involve a sedentary life style, with cyclic felling of forest as land is required, or movement of small groups of people from place to place over time, as the land becomes exhausted (Flowers et al. 1982). It can involve farming of the swamps (Dove 1980), and there are indications that the swamps at Kuk were farmed on a rotational basis at times in the past. It increasingly involves incorporation of a cash crop such as rubber (Chin 1982; Dove 1993) and at low population densities (Cramb 1989; Padoch 1988) it may be the most effective, non-destructive use of the forest, as the forest can regenerate totally. So, it can be a form of sustainable development, although there is dispute about its energy efficiency (Rambo 1980, 1984), hunting and collecting can involve less labour input.

Indonesia is one of several rapidly developing countries in Southeast Asia. It is both the largest in terms of land area and population and has a vast number of individual indigenous groups which are linguistically and culturally different, although there are inter-relations between many of them, and they are small in population size in relation to the massive number of people with a Malay origin. Its recent wealth arises from an oraganic base, not farming, but oil and natural gas from north Sumatra, and the wealth of the forests in what the Dutch called the 'Outer Islands' of which Kalimantan (Indonesian Borneo) and Sumatra are the largest. As shown in Chapter 4, forest exploitation began in prehistory but the main impact began in the 19th century on Java with felling of teak forests, to such an extent that replanting became necessary. Perhaps ironically, it is those planted trees which contain an annual record of climatic changes which will enable conclusions to be made about the impact of climatic variations on natural forest growth during this time of general clearance (cf. de Boer 1951; d'Arrigo et al. 1994; Palmer and Murphy 1993). The Dutch also protected and conserved areas of natural forest and despite their impact Java has a relatively high forest cover. They made a much smaller impression on most of the other Indonesian islands, and it was only in the 1960s, more than a decade after independence, when foreign investment was allowed, that commercial exploitation began. Of more, and continuing, importance has been the transmigration policy which has a highly political motive (Budiardjo 1986; Otten 1986; Whitten 1987), especially in turbulent regions like Irian Jaya, as well as that of relieving population pressure on Java, which it cannot do, and developing sparsely populated areas of the rest of Indonesia.

The Dutch were not the only Southeast Asian colonists to be concerned about the state of forest resources in the 19th century. In 1874 the Spanish issued a royal decree making it a crime to cut timber for commercial purposes (Makil 1984) and the Americans continued a policy of forest protection when they took over. Nor should it be forgotten that human impact on the forest has been exacerbated by the wars of the 20th century which have resulted in clearance of forest for timber exploitation to fund the war effort (Grundy-Warr 1993), to leave the enemy exposed (Mayenfeldt et al. 1978; Pfeiffer 1984; Kemf 1988) and by displaced peoples. Nevertheless the impact of peace and continued alarming rates of population growth in many areas has had a far greater impact on the rainforest than has war.

5.   The future of the rainforest

It has been stated that the history of tropical forest decline is largely unknown (Kummer 1992) but increased use of GIS techniques (cf. Brown et al. 1994) will allow greater monitoring of change in the years to come. With recent concern over global warming attention is being paid to the possible consequences of increased atmospheric CO_2 (a boon to plant metabolisms) resulting from burning, and the dispersal of aerosols. McGuffie et al. (this volume), as climatic modellers and analysts have addressed a rather different problem as an example of how their techniques can be used to understand what could happen: what would changes would take place if all the forest in the Amazon Basin, tropical Africa and Southeast Asia was physically removed ? Depending upon the assumptions used, and the information included in the dataset, different outcomes will result.   Their model suggests that the global impact might be to alter the functioning of the Hadley and Walker circulation systems through the mechanism of the upper atmospheric Rossby waves which play an important controlling role upon middle and high latitude weather.   However, as far as development of rainforest areas is concerned, the regional and local changes which might result are paramount: the dry season could be considerably lengthened, therefore shortening the natural growing season, precipitation would decrease, and would fire risk increase exacerbating land management problems. Temperature changes would be small, but diurnal variations in surface temperature, which may be important to plant growth, especially in the early stages, may increase. So, both the energy flux and hydrological balance would be considerably altered, particularly in the Amazon Basin, although less so in tropical Africa and Southeast Asia. This is not necessarily what will happen, but what **could** occur and it is a useful note on which to end this book, the note that the future is one of possibilities rather than probabilities, and, perhaps, possibilities, offer more hope for that future than do, assumed normally to be negative, probabilities.

Nevertheless, pessimists would argue that the rainforest does not have a future. Certainly the future will be very different from the past, with what survival there is, not in edaphic refuges, as possibly occurred during the phases of Quaternary climatic aridity discussed by Thomas in Chapter 2, but as a result of attempt to conserve relics of the past. Perhaps the case of what has happened in Singapore is atypical, alternatively it might be the picture of the future. In 1819 when Raffles founded modern Singapore the island was almost entirely covered in rainforest (Corlett 1992). Now less than 100 ha survives, conserved, but there are an estimated 1600 ha of tall secondary forest. So, if urban life was to suddenly disappear there is a diaspore source there which would enable the forest to recover.

We have to rely upon people realising soon the value of what is left of the rainforest. At best looking we are probably looking to a semi-urbanised parkland future, as with the few remaining forests of many temperate areas, with what little is left of the rainforest as a rather tame amenity instead of vast areas of wild land. It may be hoped that this will not prove to be the case, but at the extreme it might be claimed that something is better than nothing. The real difficulty relates not to removal of the rainforest *per se* as far as the development of such areas is concerned, but what happens afterwards. Can agriculture or agroforestry be sustained, especially if there is further massive population growth ? This is a question which cannot be readily answered. It would be very distressing to find that something of such irreplaceable beauty has been destroyed without even worthwhile material benefit resulting from that destruction, but it may be optimistically assumed that green, grassland, deserts will not be the future everywhere.

6.   Conclusion

There are many aspects of the past, present and likely future of the rainforest about which we cannot give definitive answers. Where generalisations are made they often do not measure up when looked at from an individual case perspective. While general patterns can be descerned, fractal theory is nearer to being more appropriate than Occam's razor for a consideration of the infinite intricacy of variation in the underlying physical properties of the physical base on which the trees, shrubs and herbs stand: the more we know, the more complicated it gets. The basis of fractal theory is that while discrete general patterns of variation may be recognisable, there is an underlying trend of complex variation which must be explained to reach a satisfactory explanation of, for instance, soil or plant distributions. In many incidences recently plant ecologists and zoologists have been consulted when land has been cleared in the process of resettlement schemes but geomorphological, especially, and soils advice is still not sought as widely as it should be. It would be futile to hope that most of the remaining rainforest is going to survive well into the next millenium, but it is hoped that an attempt to concentrate agricultural development on the better soils will become the norm rether than the exception, and possibly that less fertile areas will be set aside to maintain natural as opposed to bioengineered genetic diversity. All this depends upon the rate of population growth, the incidence of wars, and, most importantly, the social and political will of the peoples who occupy the areas with rainforest, and that cannot be imposed from outside, it must come from within. The rainforest may, morally, belong to the world, but some would claim that it is an increasingly amoral world, while the land the rainforest occupies does not belong to the world.

References

Anderson, A. (1994). Comment on J. Peter White's " Site 820 and the evidence for early occupation in Australia," *Quaternary Australasia*, **12** (2), 30-31.

Anderson, J.A.R. (1964). Observations on climatic damage in peat swamp forests in Sarawak, *Commonwealth Forestry Review*, **43**, 145-158.

Anderson, J.A.R. (1966). A note on two tree fires caused by lightning in Sarawak, *Malayan Forester*, **29**, 19-20.

Baillie, I.C. (1976). Further studies on drought in Sarawak, East Malaysia, *Journal of Tropical Geography*, **43**, 20-29.

Bayliss-Smith, T. and Golson, J. (1992). A Colocasian revolution in the New Guinea highlands ? Insights from Phase 4 at Kuk, *Archaeology in Oceania*, **27**, 1-21.

Beaman, R.S., Beaman, J.H., Marsh, C.W. and Woods, P.V. (1985). Drought and forest fires in Sabah in 1983, *Sabah Society Journal*, **8**, 10-30.

Borhan, M., Johari, B. and Quah, E.S. (1990). Studies on logging damage due to different methods and intensities of harvesting in hill dipterocarp forest of peninsular Malaysia, *Malaysian Forester*, **50**, 135-147.

Bowler, P.J. (1992). *The Fontana History of the Environmental Sciences*, Fontana, London.

Brooks, S.M., Richards, K.S. and Spencer, T. (1993). Tropical rain forest logging: modelling slope processes and soil erosion in Sabah, East Malaysia, *Singapore Journal of Tropical Geography*, **14** (1), 15-27.

Brown, S., Iverson, L.R. and Lugo, A.E. (1994). Land-use and biomass changes of forests in peninsular Malaysia from 1972 to 1982: a GIS approach, in V.H. Dle (ed.), *Effects of Land-use Change on Atmospheric CO_2 Concentrations: South and Southeast Asia as a case study*, Springer-Verlag, New York, pp. 117-143.

Brunig, E.F. (1964). A study of damage attributed to lightning in two areas of *Shorea albida* forest in Sarawak, *Empire Forestry Review*, **43**, 134-144.

Bruenig, E.F. (1971). On the ecological significance of drought in the equatorial wet evergreen (rain) forest of Sarawak (Borneo), in J.R. Flenley (ed.) *The Water Relations of Malesian Forests*, University of Hull, Department of Geography Miscellaneous Series No. 11, University of Hull: Hull, pp. 66-88.

Budiardjo, C. (1986). The politics of transmigration, *The Ecologist*, **16** (2/3), 111-116.

Bush, M.B., Piperno, D.R. and Colinvaux, P.A. (1989). A 6000 year history of Amazonian maize cultivation, *Nature*, **340**, 303-305.

Cannon, C.H., Peart, D.R., Leighton, M. and Kartawinata, K. (1994). The structure of lowland rainforest after selective logging in West Kalimanatan, Indonesia, *Forest Ecology and Management*, **67**, 49-68.

Chin, S.C. (1982). The significance of rubber as a cash crop in a Kenyah swidden village in Sarawak, *Federation Museum Journal*, **27**, 23-28.

Chesner, C.A., Rose, W.I., Drake, A.D.R. and Westgate, J.A. (1991). Eruptive history of Earth's largest Quaternary caldera (Toba Indonesia) clarified, *Geology*, **19**, 200-203.

Clark, J.D. and van Zinderen Bakker, E.M. (1962). Pleistocene climates and cultures in North-Eastern Angola, *Nature*, **196**, 639-642.

Clark, J.D. and van Zinderen Bakker, E.M. (1964). Prehistoric culture and Pleistocene vegetation at the Kalambo Falls, Northern Rhodesia, *Nature*, **201**, 971-975.

Clark, R. (1983) Pollen and charcoal evidence for the effects of Aboriginal burning on the vegetation of Australia, *Archaeology in Oceania*, **18**, 32-37.

Corlett, R.T. (1992). The ecological transformation of Singapore, *Journal of Biogeography*, **19**, 411-420.

Cramb, R.A. (1989). Shifting cultivation and resource degradation in Sarawak: perception and policies, *Borneo Research Bulletin*, **21** (1), 22-48.

D'Arrigo, R.A., Jacoby, G.C. and Krusic, P.J. (1994). Progress in dendroclimatic studies in Indonesia, *Terrestrial Atmospheric and Oceanic Science*, **5**, 349-363.

Dawson, A.G. (1992). *Ice age Earth: Late Quaternary geology and climate*, Routledge, London.

De Boer, H.J. (1951). Tree-ring measurements and weather fluctuations in Java from AD 1514, *Proceedings Koninklijke Nederlandse Akademie van Wetenschappen*, **B54**, 194-209.

De Selincourt, K. (1996). Demon farmers and other myths, *New Scientist*, **150** (2026), 36-39.

Dove, M.R. (1980). The swamp rice swiddens of the Kantu' of West Kalimantan, in J.I. Furtado (ed.) *Tropical Ecology and Development*, The International Society of Tropical Ecology, Kuala Lumpur, pp. 953-956.

Dove, M.R. (1993). Smallholder rubber and swidden agriculture in Borneo : a sustainable adaptation to the ecology and economy of the tropical forest, *Economic Botany*, **47**, 136-147

Flenley, J.R. (1979). *The Equatorial Rainforest: a geological history*, Butterworths, London.

Flowers, N.M., Gross, D.R., Ritter, M.L. and Werner, D.W. (1982). Variation in swidden practices in four central Brazilian Indian societies, *Human Ecology*, **10**, 203-217.

Garwood, N.C., Janos, D.P. and Brokaw, N. (1979) Earthquake-caused landslides: a major disturbance to tropical forests, *Science*, **205**, 997-999.

Glover, I.C. and Higham, C.F.W. (1996). New evidence for early rice cultivation in South, Southeast and East Asia, in D.R. Harris (ed.) *The Origins and Spread of Agriculture and Pastoralism in Eurasia*, UCL Press, London, pp 413-441.

Goldammer, J.G. and Seibert, B. (1989). Natural rainforest fires in Eastern Borneo during the Pleistocene and Holocene, *Naturwissenschaften*, **76**, 518-520.

Goldammer, J.C. and Seibert, B. (1990). The impact of drought and forest fires on tropical lowland rain forest of East Kalimantan, in J.G. Goldammer (ed.) *Fire in the Tropical Biota: ecosystem processes and global challenges*, Springer-Verlag, New York, pp. 11-31.

Gomez-Pompa, A. (1987). On Maya silviculture, *Mexican Studies*, **3**, 1-17.

Gorecki, P.P. (1986). Human occupation and agricultural development in the Papua New Guinea Highlands, *Mountain Record and Development*, **6** (1), 159-166.

Guidon, N., Pessis, A.-M., Parenti, F., Fontugue, M. and Guerin, C. (1996). Nature and age of the deposits in Pedra Furada, Brazil; reply to Meltzer, Adovasio and Dillehay, *Antiquity*, **70**, 408-421,

Grundy-Warr, C. (1993). Coexistent borderlands and intra-state conflicts in mainland Southeast Asia, *Singapore Journal of Tropical Geography*, **14**, 42-57.

Harrison, T. (1996). The palaeoecological context at Niah Cave, Sarawak: evidence from the primate fauna, *Bulletin of the Indo-Pacific Prehistory Association*, **14**, 90-100.

Hather, J.G. and Hammond, N. (1994). Ancient Maya subsistence diversity: root and tuber remains from Cuello, Belize, *Antiquity*, **68**, 330-335.

Ilife, J. (1995). *Africans: the history of a continent*, Cambridge University Press, Cambridge.

Johns, R.J. (1986). The instability of the tropical ecosystem in New Guinea, *Blumea*, **31**, 341-371.

Johns, R.J. (1988). Rainforest instability and forest management in Papua New Guinea, in J. Dargavel, K. Dixon and N. Semple (eds.), *Changing Tropical Forests: historical perspectives on today's challenges in Asia, Australasia and Oceania*, Centre for Resource and Environmental Studies, Australian National University, Canberra, pp. 103-110.

Kane, J. (1997). *Savages*, Pan Books, London.

Kemf, E. (1988) The re-greening of Vietnam, *New Scientist* 23 June, 53-58.

Kershaw, A.P. (1986) Climatic change and Aboriginal burning in north-east Australia during the last two glacial/interglacial cycles, *Nature*, **322**, 47-49

Kershaw, A.P. (1994) Site 820 and the evidence for early occupation of Australia - a response, *Quaternary Australasia*, **12** (2), 24-29.

Kershaw, A.P. (1995). The palaeoecological record from Site 820: a further response and research developments, *Quaternary Australasia*, **13** (2), 24-26.

Kummer, D.M. (1992). Measuring forest decline in the Philippines: an exercise in historiography, *Forest and Conservation History*, **36** (4), 185-189.

Longman, K.A. and Jenik, J. (1987). *Tropical Forest and its Environment*, 2nd ed., Longman, London.

Loy, T.H., Spriggs, M. and Wickler, S. (1992). Direct evidence for human use of plants 28,000 years ago: starch residues on stone artifacts from the Solomon Islands, *Antiquity*, **66**, 898-912.

Makil, P.Q. (1984). Forest management and use: Philippine policies in the seventies and beyond, *Philippine Studies*, **32**, 27-53.

Maley, J. (1996). The African rain forest - main characteristics of changes in vegetation and climate from the Upper Cretaceous to the Quaternary, *Proceedings of the Royal Society of Edinburgh*, **104B**, 31-73.

Malingreau, J.P., Stephens, G. and Fellows, L. (1985). Remote sensing of forest fires: Kalimantan and North Borneo in 1982-83, *Ambio*, **14** (6), 314-321.

Malmer, A. (1992). Water-yield changes after clear-felling tropical rainforest and establishment of forest plantation in Sabah, Malaysia, *Journal of Hydrology*, **134**, 77-94.

Maloney, B.K. (1996). *Canarium* in the Southeast Asian and Oceanic archaeobotanical and pollen records, *Antiquity*, **70**, 926-933.

Mayenfeldt. C.F.W.M. von, Noordam, D., Savenije, H.J.F., Scheltens, E.B., Torren, K. van der, Visser, P.A. and Voogd, W.B. de (1978). *Restoration of Devastated Inland Forests in South-Vietnam*, vol.1, Wageningen, Bosteelt.

Meltzer, D.J., Adovasio, J.M. and Dillehay, T.D. (1994). On a Pleistocene human occupation at Pedra Furada, Brazil. *Antiquity*, **68**, 695-714.

Otten, M. (1986). "Transmigrasi': from poverty to bare subsistence, *The Ecologist*, **16** (2/3), 71-76.

Padoch, C. (1988). Agriculture in interior Borneo: shifting cultivation and alternatives, *Expedition* **30** (1), 18-28.

Palmer, J.G. and Murphy, J.O. (1993). An extended tree ring chronology (teak) from Java, *Proceedings Koninklijke Nederlandse Akademie van Wetenschappen*, **B96** (1), 27-41.

Parenti, F., Fontugue, M. and Guerin, C. (1996). Pedra Furada in Brazil and its 'presumed' evidence: limitations and potential of the available data, *Antiquity*, **70**, 416-421.

Pfeiffer, E.W. (1984). The conservation of nature in Vietnam, *Environmental Conservation*, **11** (3), 217-221.

Primack, R. B. (1991). Logging, conservation and native rights in Sarawak forests, *Conservation Biology*,

5, 126-130.

Rambo, A.T. (1980). Fire and energy efficiency of swidden agriculture, *Asian Perspectives*, **23**, 309-316.

Rambo, A.T. (1984). No free lunch : a reexamination of energetic efficiency of swidden agriculture, in
        A.T. Rambo, and P.H. Sajise (eds.) *An Introduction to Human Ecology Research on Agricultural
        Systems in Southeast Asia,* University of the Philippines, Los Banos, pp. 154-163.

Reid, A. (1995). Humans and Forests in Pre-colonial Southeast Asia. *Environment and History*, **1**, 93-110.

Roberts, N. (1989). *The Holocene*, Blackwell, Oxford.

Stott, P.A. (1986). The spatial pattern of dry season fires in the savanna forests of Thailand, *Journal of
        Biogeography*, **13**, 104-113.

Stott, P.A. (1988). The forest as Phoenix: towards a biogeography of fire in mainland South East Asia,
        Geographical Journal, **154** (3), 337-350.

Stott, P.A. (1988). Savanna forest and seasonal fire in South East Asia, *Plants Today*, **1**, 196-200.

Stott, P.A. (1990). Stabilitiiy and stress in the savanna forest of mainland South-east Asia, *Journal of
        Biogeography*, **17**, 373-383.

Whitten, A.J. (1987). Indonesia's transmigration program and its role in the loss of tropical rain forest,
        Conservation Biology, **1** (3), 239-246.

Whitten, A.J., Damanik, S.J., Anwar, J. and Hisyam, N. (1984). *The Ecology of Sumatra*, Gadjah Mada
        University, Yogyakarta.

Wilson, S. (1985). phytolith analysis at Kuk, an early agricultural site in Papua New guinea, *Archaeology
        in Oceania*, **20**, 90-97.

Woods, P. (1989). Effects of logging, drought, and fire on structure and composition of tropical forests in
        Sabah, Malaysia, *Biotropica*, **21**, 290-298.

Dr. Bernard K. Maloney, Palaeoecology Centre, The Queen's
University, Belfast BT9 6AX, Northern Ireland

The GeoJournal Library

1. B. Currey and G. Hugo (eds.): *Famine as Geographical Phenomenon.* 1984
    ISBN 90-277-1762-1
2. S.H.U. Bowie, F.R.S. and I. Thornton (eds.): *Environmental Geochemistry and Health.* Report of the Royal Society's British National Committee for Problems of the Environment. 1985   ISBN 90-277-1879-2
3. L.A. Kosiński and K.M. Elahi (eds.): *Population Redistribution and Development in South Asia.* 1985   ISBN 90-277-1938-1
4. Y. Gradus (ed.): *Desert Development.* Man and Technology in Sparselands. 1985   ISBN 90-277-2043-6
5. F.J. Calzonetti and B.D. Solomon (eds.): *Geographical Dimensions of Energy.* 1985   ISBN 90-277-2061-4
6. J. Lundqvist, U. Lohm and M. Falkenmark (eds.): *Strategies for River Basin Management.* Environmental Integration of Land and Water in River Basin. 1985   ISBN 90-277-2111-4
7. A. Rogers and F.J. Willekens (eds.): *Migration and Settlement.* A Multiregional Comparative Study. 1986   ISBN 90-277-2119-X
8. R. Laulajainen: *Spatial Strategies in Retailing.* 1987   ISBN 90-277-2595-0
9. T.H. Lee, H.R. Linden, D.A. Dreyfus and T. Vasko (eds.): *The Methane Age.* 1988   ISBN 90-277-2745-7
10. H.J. Walker (ed.): *Artificial Structures and Shorelines.* 1988
    ISBN 90-277-2746-5
11. A. Kellerman: *Time, Space, and Society.* Geographical Societal Perspectives. 1989   ISBN 0-7923-0123-4
12. P. Fabbri (ed.): *Recreational Uses of Coastal Areas.* A Research Project of the Commission on the Coastal Environment, International Geographical Union. 1990   ISBN 0-7923-0279-6
13. L.M. Brush, M.G. Wolman and Huang Bing-Wei (eds.): *Taming the Yellow River: Silt and Floods.* Proceedings of a Bilateral Seminar on Problems in the Lower Reaches of the Yellow River, China. 1989   ISBN 0-7923-0416-0
14. J. Stillwell and H.J. Scholten (eds.): *Contemporary Research in Population Geography.* A Comparison of the United Kingdom and the Netherlands. 1990
    ISBN 0-7923-0431-4
15. M.S. Kenzer (ed.): *Applied Geography.* Issues, Questions, and Concerns. 1989   ISBN 0-7923-0438-1
16. D. Nir: *Region as a Socio-environmental System.* An Introduction to a Systemic Regional Geography. 1990   ISBN 0-7923-0516-7
17. H.J. Scholten and J.C.H. Stillwell (eds.): *Geographical Information Systems for Urban and Regional Planning.* 1990   ISBN 0-7923-0793-3
18. F.M. Brouwer, A.J. Thomas and M.J. Chadwick (eds.): *Land Use Changes in Europe.* Processes of Change, Environmental Transformations and Future Patterns. 1991   ISBN 0-7923-1099-3
19. C.J. Campbell: *The Golden Century of Oil 1950–2050.* The Depletion of a Resource. 1991   ISBN 0-7923-1442-5
20. F.M. Dieleman and S. Musterd (eds.): *The Randstad: A Research and Policy Laboratory.* 1992   ISBN 0-7923-1649-5
21. V.I. Ilyichev and V.V. Anikiev (eds.): *Oceanic and Anthropogenic Controls of Life in the Pacific Ocean.* 1992   ISBN 0-7923-1854-4

The GeoJournal Library

22. A.K. Dutt and F.J. Costa (eds.): *Perspectives on Planning and Urban Development in Belgium.* 1992   ISBN 0-7923-1885-4
23. J. Portugali: *Implicate Relations.* Society and Space in the Israeli-Palestinian Conflict. 1993   ISBN 0-7923-1886-2
24. M.J.C. de Lepper, H.J. Scholten and R.M. Stern (eds.): *The Added Value of Geographical Information Systems in Public and Environmental Health.* 1995
                                                                  ISBN 0-7923-1887-0
25. J.P. Dorian, P.A. Minakir and V.T. Borisovich (eds.): *CIS Energy and Minerals Development.* Prospects, Problems and Opportunities for International Cooperation. 1993   ISBN 0-7923-2323-8
26. P.P. Wong (ed.): *Tourism vs Environment: The Case for Coastal Areas.* 1993
                                                                  ISBN 0-7923-2404-8
27. G.B. Benko and U. Strohmayer (eds.): *Geography, History and Social Sciences.* 1995   ISBN 0-7923-2543-5
28. A. Faludi and A. der Valk: *Rule and Order. Dutch Planning Doctrine in the Twentieth Century.* 1994   ISBN 0-7923-2619-9
29. B.C. Hewitson and R.G. Crane (eds.): *Neural Nets: Applications in Geography.* 1994   ISBN 0-7923-2746-2
30. A.K. Dutt, F.J. Costa, S. Aggarwal and A.G. Noble (eds.): *The Asian City: Processes of Development, Characteristics and Planning.* 1994
                                                                  ISBN 0-7923-3135-4
31. R. Laulajainen and H.A. Stafford: *Corporate Geography.* Business Location Principles and Cases. 1995   ISBN 0-7923-3326-8
32. J. Portugali (ed.): *The Construction of Cognitive Maps.* 1996
                                                                  ISBN 0-7923-3949-5
33. E. Biagini: *Northern Ireland and Beyond.* Social and Geographical Issues. 1996   ISBN 0-7923-4046-9
34. A.K. Dutt (ed.): *Southeast Asia: A Ten Nation Region.* 1996
                                                                  ISBN 0-7923-4171-6
35. J. Settele, C. Margules, P. Poschlod and K. Henle (eds.): *Species Survival in Fragmented Landscapes.* 1996   ISBN 0-7923-4239-9
36. M. Yoshino, M. Domrös, A. Douguédroit, J. Paszynski and L.D. Nkemdirim (eds.): *Climates and Societies – A Climatological Perspective.* A Contribution on Global Change and Related Problems Prepared by the Commission on Climatology of the International Geographical Union. 1997
                                                                  ISBN 0-7923-4324-7
37. D. Borri, A. Khakee and C. Lacirignola (eds.): *Evaluating Theory-Practice and Urban-Rural Interplay in Planning.* 1997   ISBN 0-7923-4326-3
38. J.A.A. Jones, C. Liu, M-K. Woo and H-T. Kung (eds.): *Regional Hydrological Response to Climate Change.* 1996   ISBN 0-7923-4329-8
39. R. Lloyd: *Spatial Cognition.* Geographic Environments. 1997
                                                                  ISBN 0-7923-4375-1
40. I. Lyons Murphy: *The Danube: A River Basin in Transition.* 1997
                                                                  ISBN 0-7923-4558-4
41. H.J. Bruins and H. Lithwick (eds.): *The Arid Frontier.* Interactive Management of Environment and Development. 1998   ISBN 0-7923-4227-5

The GeoJournal Library

KLUWER ACADEMIC PUBLISHERS – DORDRECHT / BOSTON / LONDON